ROUTLEDGE LIBRARY EDITIONS:
COMPARATIVE URBANIZATION

Volume 5

# THE NEW HOUSING SHORTAGE

# THE NEW HOUSING SHORTAGE
## Housing Affordability in Europe and the USA

Edited by
### GRAHAM HALLETT

Routledge
Taylor & Francis Group

LONDON AND NEW YORK

First published in 1993 by Routledge

This edition first published in 2021
by Routledge
4 Park Square, Milton Park, Abingdon, Oxon OX14 4RN
605 Third Avenue, New York, NY 10017

*Routledge is an imprint of the Taylor & Francis Group, an informa business*

*British Library Cataloguing in Publication Data*
A catalogue record for this book is available from the British Library

ISBN: 978-0-367-75717-5 (Set)
ISBN: 978-1-00-317423-3 (Set) (ebk)
ISBN: 978-1-03-200383-2 (Volume 5) (hbk)
ISBN: 978-1-03-200394-8 (Volume 5) (pbk)
ISBN: 978-1-00-317391-5 (Volume 5) (ebk)

**Publisher's Note**
The publisher has gone to great lengths to ensure the quality of this reprint but points out that some imperfections in the original copies may be apparent.

**Disclaimer**
The publisher has made every effort to trace copyright holders and would welcome correspondence from those they have been unable to trace.

# The new housing shortage

Housing affordability in Europe and the USA

Edited by
Graham Hallett

London and New York

First published 1993
by Routledge
11 New Fetter Lane, London EC4P 4EE

Simultaneously published in the USA and Canada
by Routledge
29 West 35th Street, New York, NY 10001

© 1993 Graham Hallett

Typeset in Times by J&L Composition Ltd, Filey, North
Yorkshire
Printed and bound in Great Britain by
Mackays of Chatham PLC, Chatham, Kent

*British Library Cataloguing in Publication Data*
*A catalogue reference for this book is available from the*
*British Library.*

ISBN 0-415-05689-6

*Library of Congress Cataloguing in Publication Data*
*has been applied for.*

ISBN 0-415-05689-6

# Contents

vi *Contents*

# List of figures

# List of tables

# The Authors

**Professor Dr Otto Dienemann** studied building science at the Technical University of Dresden and information technology at the Humboldt University, Berlin. He worked in the building industry in the German Democratic Republic and later became Head of the Forecasting Section of the Building Academy in Berlin. In 1986 he became a Professor in the Academy and Head of the Science Division. He has been a guest Professor in various Universities and is currently a planning consultant in private practice.

**Dr Graham Hallett** has held Lectureships in Agricultural Economics and Economics in several Universities and been a Research Fellow of the Alexander von Humboldt Foundation. He is currently an Honorary Lecturer in the Centre for Housing Development and Management of the University of Wales College of Cardiff, and a freelance writer. His books include *Housing and Land Policies in West Germany and Britain; A Record of Success and Failure* (1977); *Urban Land Economics* (1979); *Housing and Land Policies in Europe and the USA; A Comparative Analysis* (1988).

**Dr E. Jay Howenstine** received his PhD from Ohio State University in 1942. He was an economist at the US Department of Agriculture, 1942–4, National Housing Administration, 1946–8, and the International Labour Organisation, 1948–67; he was International Research Coordinator, US Department of Housing and Urban Development, 1967–86. Currently member of Board of Editors, *Cities*, member of Board, Arlington Housing Corporation, and Housing Consultant. Author of numerous books and articles.

**Dr Ronald van Kempen** graduated in urban geography from the University of Amsterdam in 1985. After teaching in the Faculty of Geographical Sciences, University of Utrecht, he obtained a full-time post to study the housing situation of low-income groups in the Netherlands. He has published widely on this issue and is expanding his research to other countries in Europe.

**Professor Peter Malpass** is Professor of Housing Policy at the University of the West of England, Bristol. His main research interests are housing finance, housing management and policy implementation. He is the author of *Reshaping Housing Policy* (1990), co-author of *Housing Policy and Practice* (3rd edition 1990) and of *Implementing Housing Policy* (1992).

**Jean-Pierre Schaefer** is a graduate engineer and a Master in economics and town planning. He has worked for the building material industry (Saint Gobain), in a town planning agency (Dunkirk) and with a housing research team (*Laboratoire Logement-Nancy*). In 1977 he joined *Groupe SCIC*, the real estate branch of the *Caisse des Depots*. He is now a member of the Board of Directors of a group of companies developing owner-occupation programmes throughout France.

**Dr Rudi Ulbrich** studied economics at the University of Hamburg, and became a research assistant at the Institute for Public Finance. He then worked at the German Institute for Economic Research, Berlin. Since 1978, he has been a scientific researcher at the Institute for Housing and Environment, Darmstadt.

**Professor Dr Jan van Weesep** holds the Chair of Urban Geography and Urban Policy in the Faculty of Geographical Sciences of the University of Utrecht, and is Director of its Urban Research Program. He has published widely in the fields of housing and planning. His recently edited books include *Residential Mobility and Social Change; Studies from Poland and the Netherlands* (1990, with P. Korcelli); *Government and Housing, Studies from Seven Countries* (1990, with W. van Vliet); and *Urban Housing for the Better-Off, Gentrification in Europe* (1990, with W. Van Vliet). He is Editor-in-Chief of *Tidjschrift voor Economische en Sociale Geografie* and a member of the advisory board of *Housing Studies*.

**Dr Uwe Wullkopf** studied economics at Hamburg and became a research assistant at the School of Economics and Politics. He has been a planner at Metron in Switzerland; a visiting scholar at the University of California, Berkeley; an economist at UCLA; and a member of the professional staff of the UN, Geneva. Since 1974, he has been Director of the Institute for Housing and Environment, Darmstadt.

# Preface

Comparative international policy studies face two difficulties. To quote a comparative study of economic performance, the first is,

> the difficulty of establishing firm results; we cannot be sure what would have happened if policies had been different. . . . [The second] is the fallacious argument that, because little can be established, nothing can be learned. On the contrary, looking at different countries and at similar, if not identical, policies in different contexts can be a stimulus to our imaginations. It would be stupid to suppose that one country can transplant institutions from another – the circumstances are always subtly different – but looking at one country from the perspective of another may suggest possibilities that would not otherwise have been considered.
>
> (Graham, 1990)

It is in this spirit that my collaborators and I have written this comparative study of housing 'affordability'; the topic covers both the extreme case of homelessness and the problems of affordability, or poor accommodation, faced by people in the lower ranges of the income distribution (e.g. the bottom 20 per cent). Have these problems become worse over the past ten or fifteen years? If they have, why have they done so, and what could be done to change the situation?

Such questions are inevitably controversial. Since I am somewhat sceptical of 'value-free social science', and believe that (while being as fair and objective as possible) 'social scientists' should make their views clear, I should like to use an editor's prerogative to mention the experiences which first stimulated my interest in comparative studies of housing problems, and no doubt formed my attitudes. In the 1960s, I saw the dereliction in British private rented housing which resulted from controlled rents of ten shillings a week, and the growth of soulless, monolithic council estates, and felt that something was wrong with British housing policies. The time I spent in Canada and West Germany was therefore a revelation. In Canada there was a ready supply of private rented accommodation, at a range of prices. Canada also provided a contrast with the USA. In spite of Canada's problems (and the virtues of the USA, which I also came to appreciate) Canada did not have the huge areas of physical and social decay which blighted the cities of the richest country in the world.

West Germany seemed to me even more impressive. Not only had it achieved a remarkable recovery from 'Year zero', but it had provided housing for virtually all groups in society. There was a substantial 'social' sector, which – in marked contrast to Britain – was indistinguishable from other housing. This relative 'classlessness' of German housing reflected (as well as a less class-ridden society) a system of social housing which used a range of housing associations and even private landlords. Alongside the 'social' sector and a relatively small, but growing, owner-occupied sector was a viable private rented sector, providing housing of decent quality. What also impressed me about West Germany – in the 1960s and 1970s – was the continuity of policy; legislation was carefully discussed, both inside and outside the legislature, before being introduced, and so proved durable. In the case of the two main Acts on town planning and building regulation, ten years elapsed between the presentation of a 'draft law' and its implementation, by a Government of a different political colouring. This was a striking contrast to the 'instant legislation' of the Mother of Parliaments. Another contrast was the strength and effectiveness of German local government; the communities played an active role in the housing land market and in the supervision of housing, even though they were not directly responsible for social housing. Not everything in the West German garden was lovely. The 'cost rent' system for social housing – based on the historic cost of individual dwellings – led to serious distortions. The case for rent pooling – an admirable feature of British council housing – seemed to me so obvious that I assumed (incorrectly) that it would soon be introduced. Moreover, owner–occupancy was considerably less affordable than in Britain.

In a comparative study of housing and land policies in West Germany and Britain (Hallett 1977), I concluded that British housing policy was unlikely to be improved as long as Britain was subject to the 'U-turns' which resulted from the workings of an 'elective dictatorship' (to quote one of its supporters, Lord Hailsham) and that the fundamental need was for constitutional reform, to provide the 'checks and balances' which existed in every other developed country. However, I made some suggestions for housing policy: that housing associations should be built up alongside council housing; that tax and subsidy systems should give preferential treatment to the households in greatest housing need, rather than to any one tenure; that 'Fair Rents' should be given a rational economic interpretation, which would allow private tenancy to make a – modest, but important – contribution to housing supply. Such views were regarded, in British academia of the time, as displaying an unacceptable degree of 'Right-wing bias'.

There followed the 'Thatcher revolution'. The Conservative Governments intended to, in Mrs Thatcher's words, 'destroy socialism' (which was broadly interpreted to include elected local government, the BBC and Universities, etc.). They regarded council housing as a buttress for Labour local authorities (not altogether without justification), and were determined to eliminate it. Although I supported a more varied pattern of housing tenure, I felt that the Government's authoritarian pre-occupations were distorting the policy towards council housing. This policy was sometimes justified by reference to social housing in Western Europe, as well as to what was held up as the embodiment of free market virtues – the USA. On such points, at least, an international comparison should be able to throw some light. I began planning a comparative study of land and housing policies in North America and Western Europe which, in the event, became two studies; one of land policy in relation to housing (Hallett 1988) and the present study of housing policy with special reference to affordability.

Although the project was a purely private one, I was lucky enough to obtain the collaboration of internationally recognized experts from the various countries, who possessed practical experience as well as academic distinction. During my efforts to get the project off the ground, the dramatic changes occurred which led to German reunification. It was obvious that a chapter on the German Democratic Republic should be included. I was fortunate to obtain the collaboration of Professor Otto Dienemann, a former official of the Building Academy of the German Democratic Republic (an institution with a distinguished history going back a century before the foundation of the GDR). It has now, sadly, been largely disbanded. Professor Dienemann wrote most of his chapter in late 1990, but added a postscript in December 1991. I have left these sections as they stand; they demonstrate eloquently the hopes and disappointments of the people of the former GDR.

I am extremely grateful to the contributors for their extensive, unremunerated work, and for their forbearance in the face of editorial requests for additional material. I owe a particular debt to Jay Howenstine for his meticulous study of the chapters – not least mine – and for his suggestions for giving them more clarity and coherence. The opinions expressed in all chapters are, however, solely those of the relevant author.

## REFERENCES

Graham, A. and Seldon A. (eds) (1990) *Government and Economies in the Post-War World*, London: Routledge.

Hallett, G. (1977) *Housing and Land Policies in West Germany and Britain; A Record of Success and Failure*, London: Macmillan.

—— (ed.) (1988) *Housing and Land Policies in Europe and the USA; A Comparative Analysis*, London: Routledge.

# Acknowledgements

This book arose out of a meeting of international experts on housing policy which was financially supported by the Anglo–German Foundation for the Study of Industrial Society; the Foundation's help is gratefully acknowledged. The editor would also like to thank the Centre for Housing Development and Management of The University of Wales College of Cardiff for secretarial assistance, and the Building Societies Association, the National Association of Realtors and the Nationwide Building Society for statistical information. The editor has found the statistical compilations of Mr A.E. Holmans invaluable.

# 1 Introduction

## National similarities and differences

*Graham Hallett*

The countries in this study have many similarities in their economic
and housing systems, together with some significant differences. The
USA, UK, Netherlands, France and the 'old' German Federal
Republic are highly developed industrial economies. The German
Democratic Republic was also an industrial economy, generally
considered to be the most successful of the socialist East European
states. Because of its very different economic system, it cannot be
readily compared with the other countries, although there are some
points of comparison.

All the 'Western' countries have housing systems which are largely
market-orientated, but with substantial degrees of state intervention.
In a (pure) market economy, people get the housing they can pay
for; if they are very poor, they will get little housing, or even no
housing. This harsh rule is modified – to varying degrees in the
Western countries studied – by income-support, housing allowances,
tax concessions and the provision of subsidized housing (called
'public housing' in the USA and 'social housing' in continental
Europe and, increasingly, in the UK). In other words, in the
provision of housing, market principles are modified by public action.
In the German Democratic Republic, on the other hand, housing
was, in theory, allocated according to need, irrespective of ability to
pay; the government prided itself on having avoided the differences
in housing conditions between rich and poor which characterized the
West.

### MACRO-ECONOMIC PERFORMANCE

Economic growth, inflation and unemployment can all affect the
housing market and the housing conditions of the less well-off. The
1980s saw continued economic growth – at slightly higher rates than

## 2  The new housing shortage

*Table 1.1* Macro-economic performance

|  | Economic Growth* Average, p.a. % | | Unemployment Average, % | | Inflation Average, p.a. % | | Real Income per capita. USA = 100 |
|---|---|---|---|---|---|---|---|
|  | 1970–9 | 1980–9 | 1970–9 | 1980–90 | 1969–78 | 1978–90 | mid–1980s |
| USA | 2.0 | 2.3 | 6.2 | 7.0 | 6.6 | 6.0 | 100 |
| UK | 2.2 | 2.5 | 4.5 | 10.7 | 12.4 | 7.7 | 66.1 |
| Netherlands | 2.2 | 1.3 | 3.9 | 9.6 | 7.3 | 2.9 | 68.2 |
| France | 2.9 | 1.5 | 3.9 | 9.0 | 8.6 | 7.4 | 69.3 |
| W. Germany | 2.7 | 1.8 | 2.3 | 6.1 | 5.0 | 3.0 | 73.8 |

* *Growth of real GNP per capita*
Source: OECD, *Economic Outlook*; World Bank, *World Development Report*, 1990
p. 237.

in the 1970s in the USA and the UK, and significantly lower rates in France, West Germany and the Netherlands (Table 1.1). In terms of real income per capita, the USA, in the mid-1980s (in spite of all the talk about its relative decline) was still some 30 per cent above the average of the West European countries. In Europe, West Germany had a slight lead over the other countries, but they were grouped fairly closely together. Unemployment rose sharply in the European countries in the 1980s, compared with the previous decade. The rise in the USA was smaller and, during the 1980s, the USA had lower rates of unemployment than the UK, the Netherlands and France. Inflation in all countries was lower in the 1980s than in the 1970s, but these averages conceal some large year-to-year fluctuations.

### Housing conditions and housing systems

The countries studied can be crudely divided, according to their housing systems, into three groups. The West European countries have a good deal in common in their housing policies and tenure patterns, although the UK stands apart in some respects. The German Democratic Republic was very different, because of its comprehensive state socialism. The US housing system is at the other end of the spectrum, being predominantly 'private enterprise' – although underpinned by the Federal Government through tax relief on mortgage interest, housing allowances, and the underwriting of mortgages and of the Savings and Loan Associations.

Tenure patterns still show substantial national differences, despite some convergence. The UK has the highest levels of owner-occupation, at 67 per cent, followed by the USA (64 per cent) (Table 1.2). In the USA, higher costs of homeownership led to a fall in the percentage

*Table 1.2* Housing tenure 1988/90

|                | UK % | USA % | W. Germany % | France % | Netherlands % |
|----------------|------|-------|--------------|----------|---------------|
| Owner-occupied | 67   | 64    | 40           | 51       | 44            |
| Private rented | 7    | 33    | 45           | 26       | 13            |
| Social rented  | 26   | 3     | 15           | 23       | 43            |
| TOTAL          | 100  | 100   | 100          | 100      | 100           |

*Source:* UN. ECE and national statistics.

of owner-occupation in the 1980s (see Table 2.5). In the UK, the percentages have risen steadily; the 1987 Conservative Government set a target of 80 per cent for the year 2000. In the 'continental West European' countries, the percentages have risen but, at 40, 44 and 51 per cent, are significantly below the British level.

The differences between countries in the size of the private rented and the social rented sectors are even greater. The USA has only 3 per cent of 'public housing'. This percentage, however, underestimates the size of the 'social' sector, since there has been an expansion of a 'privatized social sector' based on 'Section 8' subsidies to private landlords. The continental European countries all have substantial social rented sectors, ranging from 43 per cent in the Netherlands to 15 per cent in West Germany (although this figure will drop sharply in the 1990s as social housing becomes deregulated). In 1980, the UK had a large social housing sector (31 per cent) which, in contrast to Continental European practice, was predominantly 'council housing'. The 'right to buy' and cut-backs in new building caused council housing to fall to 23 per cent of the housing stock in 1990. The housing association sector rose from 2 to 3 per cent of the total stock, but the whole 'social housing' sector fell to 26 per cent. The German Democratic Republic was in a different league from the other countries, with 58 per cent of the stock in 'state' and 'cooperative' housing.

In the rate of decline of its private rented sector, the UK differs from the other countries; the sector declined from 13 to 7 per cent of the total housing stock during the 1980s. The USA, the FRG and France, by contrast, all had substantial private rented sectors (32, 45, 26 per cent) at the end of the decade. The Netherlands, with 13 per cent, had significantly less. In the GDR, private ownership (both owner-occupation and private rented housing) still accounted for 41 per cent of the housing stock at the end of the decade, but the private rented sector was comprehensively state controlled, and even

*Table 1.3* Construction in the public sector as a percentage of all housing construction

|             | 1975 % | 1985 % | 1988/90 % |
|-------------|--------|--------|-----------|
| USA         | 1      | 0      | 0         |
| UK          | 52     | 21     | 14        |
| FRG         | 19     | 11     | 28        |
| GDR         | 88     | 82     | 86        |
| Netherlands | 40     | 41     | 37        |
| France      | 30     | 53     | 35        |

*Source:* ECE and national statistics

owner-occupiers were severely restricted by the state monopoly of construction and of the supply of building materials. Thus private ownership in the GDR was not comparable to private ownership in the other countries.

The levels of construction in the public social sector (as a percentage of total housing construction) differed widely, from country to country and sometimes from year to year. In the USA, during the 1980s, the figure declined from an insignificant level to zero – although this was offset by an increase in 'Section 8' grants. In the UK, where over half of construction had been in the council sector until the 1970s. the figure fell steadily, to virtually zero by 1991/2. Even taking account of the increased output of the housing associations, there was a sharp fall in 'social' construction. In France and the Netherlands, by contrast, the percentages remained in the 30–50 per cent range. In the FRG, there was a sharp decline followed by a slight recovery at the end of the decade. In the Netherlands and France, a deliberate policy was followed of increasing social sector construction when private house sales fell, so as to protect the construction industry (and households) from disruptive fluctuations. In the USA there never was a social housing programme which could be used in a counter-cyclical way. In the UK, the policy was abandoned after 1976; originally as a result of the sterling crisis of that year and, after 1979, as part of an explicit rejection of counter-cyclical policies.

## Economic changes

The 1980s saw several parallel economic developments in the countries studied. There is considerable evidence that the advanced industrial countries are moving from a system of 'organized capitalism' characterized by large firms, trade unions, stable employment and regulation to a system of 'unorganized capitalism' characterized by small firms (or loosely-organized large ones), sub-contracting, and individualistic competition. (For an optimistic review of this 'new world order' see Drucker 1989; for a more sombre review, emphasizing the potential for a growing gap between rich and poor, see Reich 1991.)

This 'system change' has affected the housing situation of the less well-off in various ways. Incomes have become more unequally distributed, and more insecure. Social housing has become more decentralized and market-orientated. Public housing services have made increasing use of the private sector. The traditional differences between housing finance and other forms of finance have (especially in the USA and UK) been eroded. As a consequence, the periodic 'rationing' of housing credit has been replaced by a 'market clearing' system in which housing finance is more readily available, but often at a higher cost, especially for low-income borrowers.

As a result of the deregulation of financial institutions (and of what the then Vice-President Bush once described as 'voodoo economics') the 1980s were a 'decade of debt' for the USA and the UK. The consequences for the housing market and the economy became clear only towards the end of the decade. In the USA, the increased indebtedness was incurred by all sectors: households, corporations and the Federal Government. The ratio of all public and private debt to GDP nearly doubled in the 1980s, to levels previously reached only in the early 1930s. In the UK, by contrast, the increase in debt was primarily confined to the household sector – where it rose even more rapidly. The continental European countries did not participate in this 'Anglo-Saxon' credit boom. In 1982, the ratios of mortgage debt to national income (GDP) were around 20 per cent in France and Germany, and just over 30 per cent in the UK and the USA; in 1989 the French and German figures were much the same, but the US percentage had risen to 45.2 and the British to 58.3.

At the end of the decade, the effects of excessive, imprudent borrowing came home to roost. In the USA, a large part of the Savings and Loan Association movement, which had been the main source of home mortgage finance, collapsed as a result of imprudent commercial investment and fraud, and is being bailed out by the Federal Government. In the UK, the banks have been seriously weakened, but the Building Societies – if not all the new 'centralized' lenders – have remained in a generally sound financial situation, as

*Table 1.4* Mortgage debt/GDP ratios

|            | 1982 % | 1989 % |
|------------|--------|--------|
| UK         | 32.1   | 58.3   |
| USA        | 33.5   | 45.2   |
| W. Germany | 22.2   | 21.9   |
| France     | 19.2   | 21.0   |

*Source:* Bank of England 1991. Table K

a result of a more effective regulatory regime and the prevalence of variable-rate mortgages. However, the burden of adjustment has been borne by recent house-buyers; there has been an unprecedented rise in repossessions.

### The Conservative counter-revolution

The profound economic changes which emerged in the 1980s (resulting largely from increased international competition and the growth of information technology) had a parallel in political thought. For some thirty years after 1945 in Western Europe, and some forty years after the New Deal in the USA there was a tendency for the role of government, in housing and other fields, to expand. In the 1980s, the pendulum swung back; the view gained increasing acceptance that direct state intervention in the supply of housing did more harm than good. In the USA and the UK the 1980s were dominated by radical conservative governments, which sought to cut public expenditure and taxes, to privatize social housing and to deregulate housing finance. In the FRG, changes were made (originally by the Social-Democrat-led Government) which are beginning to substantially reduce the supply of social housing. France and the Netherlands, on the other hand, maintained the stock of social housing, and continued to build substantial quantities of it, although less than previously. In all countries, there was a tendency to shift expenditure from social housing to personal housing allowances, and to entrust social housing to non-profit organizations, which (because of reduced subsidies) were increasingly forced to operate in a commercial manner.

The consequences of these changes for the poor and poorly-housed were hotly disputed throughout the 1980s. Supporters of privatization and deregulation argued that these changes had improved the housing situation of all groups in society and, above all, given households more independence and freedom. Critics complained that housing had become 'unaffordable' – especially for low-income households –

and that homelessness had increased. Phrases such as 'new housing shortage' and 'two-thirds society' were coined – the latter meaning that two-thirds of households were comfortable, while the rest were struggling. This continuing controversy is the background to my colleagues' accounts of housing conditions and housing policies in their respective countries.

## REFERENCES

Bank of England (1991) 'Housing Finance – An International Perspective', *Bank of England Quarterly Review*, London, February.

Drucker, P. (1989) *The New Realities*, New York: Harper Collins.

Reich, R.B. (1991) *The Work of Nations; Preparing Ourselves for 21st Century Capitalism*, New York: Knopf.

United Nations, Economic Commission for Europe (1988) *Annual Abstract of Housing Statistics*, Geneva.

# 2 The new housing shortage

## The problem of housing affordability in the United States

*E. Jay Howenstine*

The author wishes to acknowledge helpful comments from Dr Morton J. Schussheim, Senior Specialist in Housing, Congressional Research Service, US Library of Congress.

### THE SETTING

In a manner befitting a leader of the Free World, in 1949 the US Congress established as national policy 'the realization as soon as feasible of the goal of a decent home and suitable living environment for every American family'. To speed up fulfilment of this goal, in 1968 Congress adopted the Housing and Urban Development Act, which aimed to produce 26 million housing units within the next 10 years, including 6 million for low- and moderate-income families financed by federal subsidies. Accordingly, it was envisaged that by 1978 all sub-standard housing would be eliminated and all families would have a decent home and living environment.

What were the capabilities of the American economy to realize this goal? From 1946 to 1989 gross national product in 1982 dollars rose from $1.1 trillion to $4.1 trillion – almost a four-fold increase. Per capita income in 1982 dollars rose from $5,115 to $11,700, or over double (Council of Economic Advisers 1990: 296, 325, 387). In light of such an outstanding performance, there can be no reasonable doubt that the US economy had the capability to realize the goal, at least within a period of forty years. An outside observer might, therefore, well assume that by 1990 every American family would be living in a decent home and a suitable living environment.

Unquestionably, improvements in American housing conditions since World War II have been impressive. First, the quality of housing has been vastly enhanced.[1] In 1940/1, 49 per cent of the housing stock was substantially dilapidated or lacked plumbing; by 1987, sub-standard housing had fallen to 7 per cent. In 1940/1, 20 per cent of homes were overcrowded; in 1987, that had been reduced to between 2 and 7 per cent (Table 2.1).

Table 2.1 Number of households having housing problems in the United States 1987
(in 000s)

| | All households | | | Metropolitan areas | | | Central cities of metro areas | | |
|---|---|---|---|---|---|---|---|---|---|
| | Total | Owners | Renters | Total | Owners | Renters | Total | Owners | Renters |
| **Occupied units** | | | | | | | | | |
| All inadequate units | 90,884 | 58,161 | 32,723 | 70,776 | 43,549 | 27,227 | 29,984 | 14,577 | 15,407 |
| | 6,408 | 2,719 | 3,690 | 4,631 | 1,734 | 2,898 | 2,712 | 707 | 2,005 |
| Percentage | 7.1 | 4.7 | 11.3 | 6.5 | 4.0 | 10.6 | 9.0 | 4.9 | 13.0 |
| Crowded in adequate units | 1,849 | 780 | 1,069 | 1,536 | 598 | 939 | 823 | 233 | 590 |
| Percentage | 2.0 | 1.3 | 3.3 | 2.2 | 1.4 | 3.4 | 2.7 | 1.6 | 3.8 |
| Cost burden only | 18,115 | 6,120 | 11,995 | 14,751 | 4,456 | 10,295 | 7,534 | 1,588 | 5,946 |
| Percentage | 20.0 | 10.5 | 36.7 | 20.8 | 10.2 | 37.8 | 25.1 | 10.9 | 38.6 |
| Total with housing problems | 26,372 | 9,619 | 16,754 | 20,918 | 6,788 | 14,132 | 11,069 | 2,528 | 8,541 |
| Percentage | 29.0 | 16.5 | 51.2 | 29.6 | 15.6 | 51.9 | 36.9 | 17.3 | 55.4 |
| **Very low income occupied units** | 23,503 | 10,006 | 13,497 | 17,854 | 6,858 | 10,996 | 9,838 | 2,754 | 7,084 |
| All inadequate units | 3,168 | 1,103 | 2,065 | 2,290 | 666 | 1,624 | 1,514 | 315 | 1,199 |
| Percentage | 13.5 | 11.0 | 15.3 | 12.8 | 9.7 | 14.8 | 15.4 | 11.4 | 16.9 |
| Crowded in adequate units | 764 | 178 | 586 | 624 | 116 | 508 | 380 | 53 | 327 |
| Percentage | 3.3 | 1.8 | 4.3 | 3.5 | 1.7 | 4.6 | 3.9 | 1.9 | 4.6 |
| Cost burden only | 12,427 | 4,305 | 8,122 | 9,874 | 3,046 | 6,828 | 5,470 | 1,179 | 4,291 |
| Percentage | 52.9 | 43.0 | 60.2 | 55.3 | 44.4 | 62.1 | 55.6 | 42.8 | 60.6 |
| Total with housing problems | 16,359 | 5,586 | 10,773 | 12,788 | 3,828 | 8,960 | 7,364 | 1,547 | 5,817 |
| Percentage | 69.6 | 55.8 | 79.8 | 71.6 | 55.8 | 81.5 | 74.9 | 56.2 | 82.1 |

Source: Tabulations from 1987 American Housing Survey Housing and Demographic Analysis, Division U.S. Department of Housing and Urban Development

Second, the number of low and moderate-income families receiving housing subsidies has been significantly increased. According to the Department of Housing and Urban Development's definition, low-income families have an income at or below 50 per cent of the area median income, while moderate-income families have an income between 50 and 80 per cent of the area median income. The number of households directly aided by financial assistance from the Department of Housing and Urban Development (HUD) and the Farmers Home Administration (FmHA) rose from 3 million in 1977 to 5 million in 1989 (Tables 2.2, 2.3), while the total outlay of HUD and FmHA housing subsidy programme rose from $1.9 billion in 1975 to $14.9 billion in 1989 (Tables 2.3, 2.4 and Appendix 2). Third, the percentage of home-ownership rose from 44 per cent in 1938 to 65.6 per cent in 1980 (Table 2.5).

In 1989, there were 92.8 million households and 104.9 million dwelling units (including 2.9 million vacation homes), 7.7 million having been added in the previous four years (Table 2.6). Thus from a global point of view, there is a house for every household. Clearly the house building industry has chalked up some very solid achievements. It is perhaps fair to say that as a whole the American people are among the best housed in the world.

## THE PROBLEM

Alas, the good things were not to be shared by everyone. There are at least seven major dimensions to the national housing problem: the shortage of affordable housing; excessive financial burden; substandard and overcrowded accommodation; homelessness; first-time home-buyers' fading dream; the savings and loan association débâcle; and urban decay.

### Shortage of affordable housing

The National Housing Task Force Report (Rouse/Maxwell Report) attempted to estimate the gap between the supply of housing renting for under $250 a month (that is, 30 per cent of the average poverty level of household income of $10,000 in 1983) and the number of households earning $10,000 or less. It found that, from a surplus of around 4 million units in 1973, and an approximate balance in 1980, by 1983 the gap had grown to nearly 3 million (Rouse/Maxwell 1988, 6).

The Urban Institute finds that the total number of units renting for

*Table 2.2* Number of units eligible for payment in housing programmes administered by The Department of Housing and Urban Administration 1975–89

(000s)

| End of fiscal year | Gross total | Net total | Public Housing (inc. Indian Housing) | Section 8 (including vouchers) | Section 235 revised | Section 235 original | Section 256 | Rent supplement | Less Section 236/8 | Less Section 236/rent supplement | Section 202 completions |
|---|---|---|---|---|---|---|---|---|---|---|---|
| 1975 | 2,125 | NA | 1,151 | | | 408 | 400 | 165 | NA | NA | |
| 1976* | 2,397 | NA | 1,172 | 273 | | 330 | 447 | 174 | NA | NA | |
| 1977 | 2,649 | NA | 1,174 | 459 | 5 | 287 | 543 | 179 | NA | NA | |
| 1978 | 2,817 | NA | 1,173 | 666 | 12 | 249 | 544 | 171 | NA | NA | |
| 1979 | 3,031 | NA | 1,178 | 898 | 24 | 210 | 541 | 178 | NA | NA | 9 |
| 1980 | 3,268 | NA | 1,192 | 1,153 | 47 | 172 | 538 | 164 | NA | NA | 16 |
| 1981 | 3,458 | 3,297 | 1,204 | 1,318 | 82 | 158 | 537 | 157 | 95 | 66 | 27 |
| 1982 | 3,682 | 3,507 | 1,224 | 1,526 | 96 | 145 | 536 | 153 | 110 | 63 | 24 |
| 1983 | 3,840 | 3,631(a) | 1,250(a) | 1,780 | 101 | 128 | 533 | 76 | 144 | 63 | 20 |
| 1984 | 4,037 | 3,859 | 1,331 | 1,909 | 95 | 113 | 530 | 55 | 155 | 22 | 14 |
| 1985 | 4,139 | 3,943 | 1,355 | 2,010 | 88 | 111 | 527 | 45 | 169 | 27 | 15 |
| 1986 | 4,268 | 4,076 | 1,379 | 2,143 | 73 | 105 | 529 | 34 | 180 | 12 | 12 |
| 1987 | 4,340 | 4,151 | 1,390 | 2,239 | 56 | 99 | 528 | 23 | 189 | | 13 |
| 1988 | 4,429 | 4,227 | 1,397 | 2,352 | 48 | 96 | 528 | 23 | 202 | | 10 |
| 1989 | 4,512 | 4,315 | 1,403 | 2,419 | 42 | 95 | 528 | 20 | 197 | | 8 |

*Notes:* HUD adjusted housing units eligible for payment on the basis of a computer-based data source during FY 1984. The FY 1986 budget reports that adjusting to make FY 1983 figures comparable with those of FY 1984 raises the 1983 public housing figure to 1,313,816, the gross figure for all assisted housing to 3,903,880 and the net to 3,737,144.
* Includes TQ adjustment for change in length of fiscal year.
NA Not available.
*Source:* HUD Summary Budgets, FYs 1977–91.
For brief descriptions of various HUD programmes, see Explanatory Notes at the end of Table 2.6.

*Source:* Grace Milgram (1990) *Trends in Funding and Numbers of Households in HUD-assisted Housing, Fiscal Years 1975–1990*, Washington DC, Congressional Research Service, Library of Congress: 90–266 E: 19.

Table 2.3 Farmers' Home Administration Assistance Programmes 1975–90

| Year | Number of families (000s) | | | | Loan value ($ million) | | | | Subsidy ($m) (covers loss of Rural Housing Insurance Fund) |
| | Rental housing (Section 515) | | Home-ownership (Section 502) | | Rental housing (Section 515) | | Home-ownership (Section 502) | | |
| | Annual addition | Housing stock | Annual addition | Housing stock | Annual addition | Total outstanding | Annual addition | Total outstanding | |
| | (1) | (2) | (3) | (4) | (5) | (6) | (7) | (8) | (9) |
|---|---|---|---|---|---|---|---|---|---|
| 1975 | 17.4 | 47.5 | 56.4 | – | 243 | – | 1,186 | – | 175 |
| 1976 | 25.0 | 72.5 | 60.4 | – | 198 | – | 1,356 | – | 327 |
| 1977 | 25.0 | 97.5 | 61.0 | 258.2 | 513 | – | 1,486 | 4,982 | 320 |
| 1978 | 29.5 | 127.0 | 62.8 | 285.0 | 600 | – | 1,720 | 5,974 | 322 |
| 1979 | 34.0 | 161.0 | 64.3 | 310.1 | 822 | – | 2,007 | 7,089 | 504 |
| 1980 | 27.5 | 188.5 | 62.7 | 326.0 | 825 | – | 2,220 | 8,009 | 653 |
| 1981 | 24.5 | 213.0 | 61.5 | 344.9 | 842 | – | 2,328 | 9,174 | 1,110 |
| 1982 | 25.5 | 238.5 | 56.9 | 364.8 | 954 | – | 2,300 | 10,362 | 1,508 |
| 1983 | 20.0 | 258.5 | 51.2 | 404.5 | 802 | – | 2,112 | 11,994 | 1,643 |
| 1984 | 22.5 | 281.0 | 43.5 | 396.5 | 919 | – | 1,836 | 12,150 | 2,136 |
| 1985 | 23.5 | 304.5 | 40.8 | 400.7 | 903 | – | 1,781 | 12,661 | 2,715 |
| 1986 | 18.5 | 323.0 | 25.5 | 395.6 | 652 | – | 1,147 | 12,704 | 2,964 |
| 1987 | 14.5 | 337.5 | 24.1 | 383.0 | 555 | – | 1,133 | 12,405 | 3,660 |
| 1988 | 14.5 | 352.0 | 25.9 | 370.2 | 555 | – | 1,252 | 12,216 | 2,678 |
| 1989 | 13.5 | 365.5 | 24.8 | 364.0 | 555 | 8,885 | 1,246 | 12,243 | 2,667 |
| 1990 | 13.5 | 379.0 | – | 356.7 | 572 | 9,430 | – | – | – |

Source: Columns (1) (2) (5) (6): Larry Anderson, Multi-housing Servicing and Property Management, FmHA
Columns (3) (4) (7) (8): David Mead, Management Analyst, Information Resources Division, FmHA
Column (9): Don Brooks, Senior Analyst, Budget Staff, FmHA

Table 2.4 Outlays in housing assistance programmes administered by HUD 1975–89
(in millions of dollars)

| Fiscal Year | Total assisted housing (excl. college housing) | Public housing (excl. operating subsidy) | Section 8 (including vouchers) | Section 235 | | | Section 236 | Rent Supplement | Section 202[a] | Public housing operating subsidy |
|---|---|---|---|---|---|---|---|---|---|---|
| | | | | Restructured | Revised | Original | | | | |
| 1975 | 1,732 | 971 | — | — | — | 194 | 391 | 174 | −12 | 339 |
| 1976+TQ | 2,453 | 1,270 | 75 | — | — | 201 | 642 | 263 | −17 | 632 |
| 1977 | 2,405 | 1,073 | 367 | — | 2 | 125 | 585 | 251 | 3 | 522 |
| 1978 | 2,896 | 1,073 | 847 | — | 7 | 99 | 616 | 252 | 176 | 695 |
| 1979 | 3,537 | 1,163 | 1,371 | — | 17 | 81 | 638 | 265 | 459 | 654 |
| 1980 | 4,506 | 1,360 | 2,104 | — | 44 | 70 | 656 | 271 | 752 | 824 |
| 1981 | 5,726 | 1,472 | 3,115 | — | 110 | 86 | 665 | 277 | 817 | 928 |
| 1982 | 6,859 | 1,766 | 4,085 | — | 152 | 106 | 669 | 279 | 742 | 1,007 |
| 1983 | 7,766 | 1,664 | 4,994 | — | 166 | 115 | 637 | 187 | 799 | 1,541 |
| 1984 | 8,753 | 1,686 | 6,030 | — | 180 | 89 | 657 | 109 | 661 | 1,135 |
| 1985 | 9,973 | 2,202 | 6,818 | 9 | 175 | 82 | 619 | 65 | 501 | 1,205 |
| 1986 | 10,020 | 1,700 | 7,430 | −2 | 144 | 67 | 633 | 46 | 522 | 1,180 |
| 1987 | 9,765 | 773 | 8,125 | 10 | 113 | 58 | 637 | 47 | 376 | 1,388 |
| 1988 | 11,021 | 1,037 | 9,133 | 7 | 112 | 56 | 628 | 46 | 297 | 1,488 |
| 1989 | 12,215 | 1,524 | 9,917 | 6 | 72 | 36 | 611 | 48 | 356 | 1,519 |

[a] Net loans (loans minus repayments and interest payments)

*Source:* HUD Summary Budgets, FY 1977–FY 1991
For brief descriptions of various HUD programmes, see explanatory notes at end of Table 2.6.

*Source:* Grace Milgram (1990) *Trends in Funding and Numbers of Households in HUD-Assisted Housing, Fiscal Years 1975–1990,* Washington DC, Congressional Research Service, Library of Congress: 90–266E: 17.

*Table 2.5* Home-ownership rates by selected age groups 1938–89

|  | 1938 | 1973 | 1980 | 1983 | 1987 | 1988 | 1989 |
|---|---|---|---|---|---|---|---|
|  |  |  |  | *per cent* |  |  |  |
| All households | 44 | 64.4 | 65.6 | 64.9 | 64.0 | 63.9 | 63.8 |
| Head under 25 | 23 | 21 | 19 | 16 | 15 | n.a. |  |
| Head 25–34 | 51 | 52 | 47 | 45 | 45 | n.a. |  |
| 35–44 | 70 | 72 | 69 | 66 | 67 | n.a. |  |
| 45–64 | 75 | 78 | 78 | 78 | 77 | n.a. |  |
| 65 or over | 69 | 72 | 74 | 75 | 75 | n.a. |  |

*Source:* Joint Center for Housing Studies of Harvard University, as tabulated from HUD *Annual Housing Survey* 1973 and 1980, and US Dept. of Commerce *Current Population Survey* 1983, 1987, 1988, 1989. The 1938 figure from National Housing Task Force, *A Decent Place to Live*, Washington DC (1988) p. 4.

*Source:* Barbara L. Miles (1990) *Housing Policy: Homeownership Affordability*, Washington DC, Congressional Research Service, Library of Congress, Co. IB88108: 3.

less than $300 (in 1988 dollars) declined from just over 10 million in 1974 to 8.5 million in 1985. As a result, the number of poor renters' households is now about three times as high as the number of low-cost affordable units (Turner 1990, 31). Another study by the Neighborhood Reinvestment Corporation has estimated that the gap will rise from a nation-wide level of about 3.25 million units in 1983 and 6.8 million in 1993 to 10.3 million in the year 2000 (Clay 1987, 4).

### Excessive financial burden

Excessive financial burden is measured by the number of households which must pay more than a reasonable percentage of their income for housing. Traditionally 25 per cent has been regarded as a reasonable proportion of family income to spend on housing. In 1981, however, the percentage was raised to 30 per cent. This change was introduced not after a considerable analysis of what constituted 'reasonableness' in housing burden, but simply as a convenient means of dealing with a political crisis which otherwise would have required a larger appropriation for federal housing subsidies.

    Great dissatisfaction has been expressed with this rather arbitrary concept of reasonableness, in fact the Rouse/Maxwell Report recommends a more flexible definition of reasonable rent burden providing for a sliding scale based on family size and income. That is, larger families with lower incomes would pay a smaller proportion of income, while small, relatively better-off households would pay a larger proportion (Rouse/Maxwell 1988, 44). On the other hand, households with incomes of $40,000 or more pay less than 15 per cent of their income for housing costs (Table 2.7). In short, the operating principle seems to be, the less income a family earns, the more it can expect to pay for housing – which from the standpoint of social need should be quite the other way round.

By contrast, in European countries, the average pre-World War II

*Table 2.6* Household units 1940–89 and housing inventory 1965–89 (millions)

| Year | Households | Total | Housing inventory Seasonal vacant (2nd home) | | Total occupied Owner | Renter |
|------|-----------|-------|------------------|-------|------|--------|
| 1940 | 34.9 | – | – | – | – | – |
| 1950 | 43.6 | – | – | – | – | – |
| 1955 | 47.9 | – | – | – | – | – |
| 1960 | 52.8 | – | – | – | – | – |
| 1965 | 57.4 | 64.2 | 1.7 | 6.7 | 36.2 | 21.3 |
| 1966 | 58.4 | 65.2 | 1.8 | 6.7 | 37.1 | 21.4 |
| 1967 | 59.2 | 66.0 | 1.8 | 6.5 | 37.8 | 21.6 |
| 1968 | 60.8 | 67.2 | 1.7 | 6.2 | 38.9 | 22.0 |
| 1969 | 62.2 | 68.5 | 1.8 | 6.2 | 40.0 | 22.2 |
| 1970 | 63.4 | 69.8 | 1.7 | 6.1 | 40.8 | 22.8 |
| 1971 | 64.8 | 71.3 | 1.7 | 6.2 | 41.8 | 23.3 |
| 1972 | 66.7 | 73.3 | 1.7 | 6.4 | 43.1 | 23.8 |
| 1973 | 68.3 | 75.4 | 1.7 | 6.6 | 44.4 | 24.4 |
| 1975 | 71.1 | 78.8 | 1.7 | 6.9 | 46.5 | 25.5 |
| 1976 | 72.9 | 80.2 | 1.6 | 6.8 | 47.5 | 25.9 |
| 1977 | 74.1 | 81.6 | 1.6 | 6.9 | 48.5 | 26.3 |
| 1978 | 76.0 | 83.5 | 1.7 | 6.9 | 49.7 | 26.8 |
| 1979 | 77.3 | 85.1 | 1.7 | 7.1 | 51.1 | 26.8 |
| 1980 | 80.8 | 86.7 | 2.1 | 8.1 | 52.2 | 27.4 |
| 1981 | 82.3 | 90.9 | 1.9 | 8.1 | 54.1 | 28.7 |
| 1982 | 83.5 | 91.9 | 1.8 | 8.1 | 54.2 | 29.5 |
| 1983 | 83.9 | 92.0 | 1.8 | 8.5 | 54.7 | 29.9 |
| 1984 | 85.3 | 95.3 | 2.0 | 8.9 | 55.7 | 30.7 |
| 1985 | 86.8 | 97.3 | 2.0 | 9.4 | 56.2 | 31.7 |
| 1986 | 88.5 | 99.3 | 2.4 | 10.2 | 56.8 | 32.3 |
| 1987 | 89.5 | 101.8 | 3.0 | 11.3 | 57.9 | 32.6 |
| 1988 | 91.1 | 103.7 | 3.1 | 11.6 | 58.7 | 33.2 |
| 1989 | 92.8 | 105.0 | 3.0 | 11.5 | 59.8 | 33.7 |

*Source:* Household data – Bureau of Census, *Households, Families, Marital Status, and Living Arrangements: March 1989 (Advance Report)*, Washington DC: Government Printing Office, 1989, Current Population Reports, Series P-1-: 441: 6–9; Housing Inventory data – Bureau of Census, Table extracted, *Current Population Survey/Housing Vacancy Survey, 1989*, Series H-111, supplied by Wallace Fraser, Bureau of Census analyst, July 16 1990

rent-to-income ratio was around 10 per cent, and in the housing allowance programmes adopted after World War II, the effective ratios averaged between 10 and 15 per cent, with many of them including sliding scales to take account of the size of the family and its income. In Austria, France and the United Kingdom, for very low levels of income the ratio is zero (Howenstine 1986, 25–8).

An alternative method of determining 'affordability' is the so-called 'market-basket' concept. This approach determines first the cost of the household's other basic necessities, using the Bureau of Labor Statistics 'urban family budgets', and then finds out how much income is left for housing costs. In this system, the '30 per cent of income' standard is not at all appropriate for many low-income households. For example, under the 30 per cent standard in 1985, a family of two with an income of $6,000 could afford to pay $1,800 a year for housing, but under the market-basket approach, the family could not fully meet the costs of other necessities and still afford anything for housing. The result is that the number of large renter households with housing affordability problems is substantially greater under the market-basket approach than under the 30 per cent rule. Conversely, the number of small renter households with affordability problems is less under the market-basket approach than under the 30 per cent-of-income standard. Overall, the total number of renter households with affordability problems is roughly the same under both approaches (Leonard 1989, 69–72).

Nevertheless, taking the 30 per cent rent-to-income ratio as the official measure of affordability, in 1987, 18.1 million households, or 20 per cent of total households were shouldered with a cost burden (Table 2.1). Moreover, among very-low-income renters, i.e. those below 50 per cent of area of median income, 74 per cent spend more than 30 per cent of their incomes on rent, while 47 per cent spend more than half of their income on housing (Table 2.8).

### Sub-standard and overcrowded accommodation

Although, as noted above, great progress has been achieved in upgrading the quality of the nation's housing stock, significant deficiencies remain. According to the 1987 American Housing Survey, 7.1 per cent of the housing stock, or 6.4 million units, failed to meet minimum housing standards, the principal factors being deficient plumbing and kitchen facilities (Table 2.1).

According to HUD standards, a housing unit is considered over-crowded if it houses more than one person per room. In 1987, 2 per

*Table 2.7* Monthly housing costs as a per cent of income for various levels of income 1987
(Based on median per cent within each income level)

| Levels of income | Per cent of income |
|---|---|
| $1 to $4,999 | 70 |
| $5,000 to $9,999 | 40 |
| $10,000 to $14,999 | 30 |
| $15,000 to $19,999 | 24 |
| $20,000 to $29,999 | 21 |
| $30,000 to 39,999 | 18 |
| $40,000 to $59,999 | 15 |
| $60,000 to $79,999 | 13 |
| $80,000 to $99,999 | 11 |
| $100,000 to $119,999 | 10 |
| $120,000 or more | 7 |

*Source:* Bureau of Census, *American Housing Survey for the US: 1987*, Washington DC, Government Printing Office 1989, Table 2–20: 73

cent of the households living in adequate (i.e. standard) housing, or 1.8 million families, lived in crowded conditions (Table 2.1). Because of statistical limitations, there is no way of knowing how many families living in sub-standard housing conditions are also over-crowded, but it is reasonable to believe the proportion would be considerable. Therefore, one can only conclude that between 2 and 7 per cent of the nation's households live in overcrowded conditions.

In addition to overcrowding, there has also been a substantial increase in 'doubling up' i.e. sharing the house unit with relatives or non-relatives. Although this 'doubling up' may not necessarily be overcrowding in the statistical sense, it may definitely be overcrowd-ing in the social sense. In 1978, 15 per cent of poor households doubled up in some way; in 1987 this had increased to 23 per cent (HUD 1989a, 52).

**Homelessness**

Outside the housing market are the homeless. Langley Keyes dis-tinguishes three types of homelessness. First, there is economic homelessness – people who are homeless simply because they cannot afford a place to live. The second is situational homelessness – people who are victims of abuse, who find themselves in some kind of transitional situation. Although many will require various types of support and training services to regenerate their motivation, they do not need permanent help. Third, there are the chronic homeless

*Table 2.8* Number of households paying 30 per cent and 50 per cent of their incomes for housing costs 1975–87
(units in thousands)

|  | 1975 | 1981 | 1985 | 1987 Prelim |
|---|---|---|---|---|
| **Total** | | | | |
| Paying 30%+ | 10,664 | 15,554 | 20,588 | 20,895 |
| (% of total) | (14.7) | (18.7) | (23.3) | (23.0) |
| Paying 50%+ | | 7,498 | 9,347 | 9,297 |
| (% of total) | | (9.0) | (10.6) | (10.2) |
| **Owners total*** | | | | |
| Paying 30%+ | 2,871 | 4,672 | 6,938 | 6,707 |
| (% of total) | (6.1) | (8.6) | (12.4) | (11.5) |
| Paying 50%+ | | 2,096 | 2,691 | 2,561 |
| (% of total) | | (3.9) | (4.8) | (4.4) |
| **Very low income** | | | | |
| Paying 30%+ | 2,445 | 3,413 | 4,722 | 4,821 |
| (% of total) | (27.2) | (34.1) | (50.4) | (48.2) |
| Paying 50%+ | | 1,837 | 2,288 | 2,288 |
| (% of total) | | (18.4) | (24.4) | (22.3) |
| **Renters total** | | | | |
| Paying 30%+ | 7,793 | 10,882 | 13,650 | 14,188 |
| (% of total) | (30.4) | (37.7) | (42.3) | (43.3) |
| Paying 50%+ | | 5,402 | 6,656 | 6,736 |
| (% of total) | | (18.7) | (20.6) | (20.6) |
| **Very low income** | | | | |
| Paying 30%+ | 6,395 | 8,077 | 9,347 | 9,988 |
| (% of total) | (63.2) | (68.4) | (71.9) | (74.0) |
| Paying 50%+ | | 5,090 | 6,188 | 6,298 |
| (% of total) | | (43.1) | (47.6) | (46.7) |

* Owners with mortgage have a payment standard of an additional 10%
*Sources: Annual Housing Survey, 1975–83; American Housing Survey, 1985–87;*
Housing and Demographic Analysis Division, Office of Policy Development and Research, Department of Housing and Urban Development, 1989

– the drug abusers, the alcoholics, the chronically mentally ill – a residential category that has always existed. They need permanent housing, often of an institutionalized nature (General Accounting Office 1989, 57).

Homelessness is probably the most dramatic evidence of the low-cost housing shortage. In the early 1980s relatively few people lived on the streets or in homeless shelters; and most of those who did were single men, many of whom were alcoholics and drug addicts. By the late 1980s, the size and composition of the homeless had changed markedly (Burt and Cohen 1989).

There is much controversy about the size of the homeless population with estimates varying from 350,000 (HUD 1984) to 3 million (National Coalition for the Homeless) (Zigas 1989, 3). More recently, the Urban Institute concluded that 'cautious estimates' place the number between 500,000 and 600,000, about 15 per cent of whom are children. One of the most disturbing elements in the contemporary situation is the increasing number of the new homeless that are families with children (Turner and Reed 1990, 2).

### First-time home-buyers' fading dream

Home-ownership is one of the time honoured pillars of the American value system. It provides a feeling of security, it is a powerful motivating factor behind a strong work ethic, and it is generally considered to be the preferred environment in which to raise children.

The 1980s, however, have been a turbulent and discouraging period for first-time home buyers, particularly for those between the ages of 25 and 35. As a result of accelerating home prices and high interest rates, many young prospective home owners – in the absence of assistance from parents and relatives – do not have the savings and income required to make the down payment and cover the monthly charges for home-ownership. In 1949, the average 30 years-old home-owner devoted 14 per cent of the monthly pay check to mortgage payments; by 1985 the figure had risen to 44 per cent (Dreier 1988, 2–3). As a result the Rouse/Maxwell Report estimates that approximately 2 million young American families have been priced out of the market (Rouse/Maxwell 1988, 8) causing the percentage of home-owners to fall from a peak of 65.6 per cent in 1980 to 63.8 per cent in 1989 (Table 2.5).

### The savings and loan association débâcle

For most of this century, savings and loan associations (S and Ls) have been the main vehicle for financing home-ownership. With the legal right to pay one-half per cent higher interest to depositors than commercial banks, the S and Ls were guaranteed a protected supply of capital. However, in the late 1970s trouble began to brew.

Basically the crisis arose because, in the new era of deregulation beginning in 1980, the interest rate that banks had to pay depositors skyrocketed – to between 12 and 14 per cent – while the rate they received from the bulk of their mortgage loans remained fixed – at around 5 to 6 per cent. Thus during the period of peak interest rates,

1981–2, already the total unrealized portfolio loss of all S and Ls was more than $150 billion (Rosen 1984, 163). Although interest rates declined in the later 1980s, the new easy-going atmosphere of deregulation and competition with commercial banks opened the door not only to new high risk lending, but also to new opportunities for fraudulent behaviour.

By 1989, the teetering S and L system finally collapsed, and hundreds of institutions went bankrupt. When the Financial Institution Reform, Recovery and Enforcement Act was passed in August 1989, the 'official' estimated cost to the taxpayer was around $150 billion. However, with costs continuing to rise monthly, as more and more bad loans and fraudulent behaviour were uncovered, William Seidman – the Chairman of the Federal Deposit Insurance Corporation and the key man in the bailout – stated that an estimated figure between $300 and $500 billion was more likely (Fricker and Pizzo 1990).

The future of the S and L system depends on many factors beyond the scope of this review, such as the level of national economic prosperity, national monetary and fiscal policy, and the general policy of deregulation of capital markets. It may well be that the housing sector will not regain the protected status it enjoyed under the traditional S and L system (Rosen 1984; Downs 1983b).

**Urban decay**

The worth of a home is not a self-determined value; it is very much tied up with the quality of its surroundings. One of the distressing aspects of big city life is that frequently an otherwise structurally sound housing unit is overwhelmed by a deteriorating neighbourhood characterised e.g. by absentee ownership, housing abandonment, arson, vandalism, high crime rates, drug traffic, mortgage and tax foreclosures, poor maintenance of public facilities, noisy traffic and general environmental degradation. Moreover, in inner cities, the existing disproportionate concentration of the poor often burdens the community with above-average unemployment, broken families, ill health, social pathologies, capital disinvestment, and a general loss of self-esteem, social status and confidence in the future.

As a result, deteriorating neighbourhoods, which thrive on dependency and the underground economy, tend to be perpetuated with negative attitudes toward work and family life. In short, translating the concept of a 'decent home' into reality for the urban poor goes far beyond the confines of a narrowly defined national housing policy.

There is hardly a large metropolitan area in the US that does not have a serious problem of urban decay.

In view of the above litany of problems, what in Heaven's name has gone wrong?

## THE CAUSES OF THE PROBLEM

### Poor housing

At least five major factors have contributed to the housing affordability problem in contemporary America: (1) poverty and low income, (2) rising housing costs, (3) shrinkage in the supply of affordable rental housing, (4) shortcomings of the public housing programme, and (5) existing housing subsidies going to the wrong people.

*Poverty and low income*

Basically, the inability of low-income families to find decent housing they can afford is attributable to the fact that they do not have enough income, that they are poor.

There are three kinds of poor: the working poor; the unemployed; and the unemployable. In 1986, one quarter of all full-time jobs, 24 million, did not pay enough to raise a family of four above the poverty line. Of the 241 million people in the US in 1986, one in seven – or over 32 million – lived below the poverty line as defined by the Department of Labor ($11,203 for a family of four; $5,572 for an individual). Almost one-third of all households earned less than $15,000 (Rouse/Maxwell 1988, 5). Among blacks, almost one in three persons were below the poverty line (Council of Economic Advisors 1990, 328). The majority of poor people are renters. In 1987, 63 per cent of all poverty level households were living in rental housing (Rouse/Maxwell 1988, 5).

*Rising housing costs*

Housing costs have risen more rapidly than family incomes. Between 1975 and 1988, real rent levels increased by 17 per cent, while real incomes among renters actually fell by 4 per cent (Turner and Reed 1990, 2). During the period 1974–88, for single parent households with heads aged 25 to 34, real income declined by 9 per cent, and for those heads under 25 the fall was 15 per cent. At the same time the number of renters with income at or below the poverty line increased by 50 per cent in the 1974–88 period (Apgar *et al.* 1989, 8). Furthermore, the average cost in real terms of rehabilitating a rental housing unit, which remained roughly constant from 1970 to 1982, more than doubled between 1982 and 1986 (Rouse/Maxwell 1988, 6). The resulting upgrading of rental properties that were once affordable often put that housing out of the reach of the poor.

Ownership costs also show a long-term increase. The annual after-tax cash cost of ownership in 1988 dollars for a first-time home buyer of a modest single family house rose from 18.3 per cent of income in 1967 to a peak of 37.4 per cent in 1982, after which it declined to 33 per cent in 1988. High home-ownership prices force potential first-time home-buyers to remain in the rental housing market, bidding up the price of rental housing. This in turn limits further the home-ownership opportunities of young renters, as increased rents reduce the ability to save for a home down-payment.

*Shrinkage in the supply of affordable rental housing*

The 'trickle down' process is a well recognized concept in the housing market. As a household builds and occupies a new house, existing housing is vacated, making it available for occupancy by lower income households, who in turn liberate their own accommodation as they move up-market. Through this continuous chain of moves, more housing is ultimately made available for the poor. Unfortunately in recent years, the trickle down process has slowed considerably.

The actual physical shrinkage of affordable housing units is one of the dismaying features of the American housing market. From 1973 to 1983, 4.5 million privately owned housing units were permanently removed from the housing stock either through demolitions or structural conversions (Rouse/Maxwell 1988, 6). In addition a large number of affordable housing units were taken out of the rental stock through conversion to condominium ownership. While no nation-wide statistics are available, in Arlington County, Virginia, from 1980 to 1990 there was a loss of affordable housing units to condominium conversion of 6,700 in total housing stock of 83,000 or a loss of 8 per cent (Arlington County 1990, 11).

It might be thought that new public and/or private construction could be expanded to compensate for this shrinkage in the rental housing stock, but this offers little immediate promise. As regards new public construction, the public housing programme has never played the major role in providing affordable housing in the US that

*Table 2.9* Public housing completions 1939–91

| Year | Units | Year | Units |
|------|-------|------|-------|
| 1939 | 4,960 | 1964 | 24,488 |
| 1940 | 34,308 | 1965 | 30,769 |
| 1941 | 61,065 | 1966 | 31,483 |
| 1942 | 36,172 | 1967 | 38,756 |
| 1943 | 24,296 | 1968 | 72,638 |
| 1944 | 3,269 | 1969 | 78,003 |
| 1945 | 3,080 | 1970 | 73,723 |
| 1946 | 1,925 | 1971 | 91,539 |
| 1947 | 466 | 1972 | 58,590 |
| 1948 | 1,348 | 1973 | 52,791 |
| 1949 | 547 | 1974 | 43,928 |
| 1950 | 1,255 | 1975 | 24,514 |
| 1951 | 10,246 | 1976 | 6,862 |
| 1952 | 58,258 | 1977 | 6,229 |
| 1953 | 58,214 | 1978 | 10,295 |
| 1954 | 44,293 | 1979 | 44,019 |
| 1955 | 20,899 | 1980 | 15,109 |
| 1956 | 11,993 | 1981 | 33,631 |
| 1957 | 10,513 | 1982 | 28,529 |
| 1958 | 15,472 | 1983 | 27,876 |
| 1959 | 21,939 | 1984 | 24,092 |
| 1960 | 16,401 | 1985 | 19,267 |
| 1961 | 20,565 | 1986 | 15,464 |
| 1962 | 28,682 | 1987 | 10,415 |
| 1963 | 27,327 | 1988 | 9,146 |
| | | 1989 | 5,238 |
| | | 1990 | 6,677 (EST) |
| | | 1991 | 9,174 (EST) |

*Sources:* The Report of the President's Committee on Urban Housing, 1968, 61; HUD 1980b, 204; and Special Memorandum, Low Income Housing Information Service March 1989 and February 1990. Rachel G. Bratt (1989) *Rebuilding a Low-Income Housing Policy*, Philadelphia, Temple University Press, p. 57

it has in most European countries. Inaugurated only in 1937, the public housing stock grew slowly (Table 2.9) until it reached 1.4 million in 1989, or approximately 1.3 per cent of the nation's housing stock (Tables 2.2, 2.6). During the 1980s, the Reagan Administration viewed the public housing programme with disfavour.

As regards new private rental housing, from 1976 to 1982, over a million new federally subsidized, privately-owned units were added to the supply. In recent years, however, fewer than 25,000 subsidized, privately-owned units have been built annually. Without new financial incentives, the private sector cannot be expected to produce much housing for low-income households. In 1986, only 7.5 per cent

(30,000 units) of the private sector's new unsubsidized multi-family production rented for less than $300 a month (Rouse/Maxwell 1988, 6).

*Failings of the public housing programme*

One factor that has contributed to the impasse in trying to close the gap between the supply and demand of low-cost, decent housing has, perhaps surprisingly, been the failings of the public housing programme. The origins of these failings are in many ways attributable to outside forces, but in considerable measure they have arisen within the programme itself. To begin with, the New Deal public housing programme was conceived in 1937 mainly as a means of attacking unemployment and of providing jobs, and only secondarily as a means of meeting people's housing needs. In fact, the programme was viewed as providing temporary housing for the 'deserving poor', who could not find decent housing on the private market, but not for those with little or no means to pay rent (i.e. the 'undeserving' (?) poor).

During its early history, the public housing programme was strongly opposed by real estate interests – the builders, the developers and the mortgage bankers. As a result of this opposition and the principle of decentralized political control over housing policy (in which the local power structure was often not particularly sympathetic to the problems of the poor), public housing had to be different from, and could not compete with, private housing. Moreover, it ran into project design and site deficiencies (e.g. project size and lack of defensible space), that tended to make it racially segregated. Although during the last three decades the real estate lobbies reversed their position and became active supporters of most national housing subsidy programmes, the presence of powerful producer interests, not only in the hammering out of basic legislative principles but also in the administration of HUD programmes, has not been altogether compatible with consumer interests, especially those of low- and moderate-income families (Bratt *et al.* 1989, 55–67).

After World War II, when public housing became more accepted as a means of providing permanent shelter for the poor generally, public housing authorities were required by law to evict families, when their income rose above original eligibility requirements. This was to ensure the application of the laudable social principle that federal housing subsidies should go to the most needy. Consequently, there was a growing concentration of the lowest income families in public housing projects, with an increasing proportion of single-parent families on welfare. Moreover, in practice, racial segregation in high multi-storey buildings sometimes created ghettos where crime, drugs and violence thrived. Thus American public housing forfeited the opportunity of developing stable, economically-mixed communities, such as developed in many European countries, where from 20 to 50 per cent of the occupants have had incomes above original entry-requirements (Howenstine 1986, 137).

Although the above failings have by no means characterized a majority of public housing projects (only one quarter of all units are found in large inner city areas), the stereotype became so well entrenched in political dialogue that it has been difficult to maintain wide popular support for the public housing programme.

## Housing subsidies going to the wrong people

After the federal income tax code was set up in 1913, the law provided that home owners could deduct mortgage interest payments and local property taxes from their gross incomes in calculating their income taxes. The US Treasury calls such deductions 'tax expenditures'. These subsidies were conceived as an important incentive for promoting widespread home-ownership – a *summum bonum* in American society. And indeed, they do have a special role to play for low- to middle-income families, who might not otherwise be willing or able to afford the costs of home-ownership.

Initially these subsidies were relatively modest. However, as a result of rapidly rising residential property values, particularly in the last decade, home-ownership (tax expenditure) subsidies in 1991 reached an estimated $59.6 billion – $46.6 billion for mortgage interest payments and $12.4 billion for local property taxes – over three times the size of traditional housing subsidies for the poor (Appendix 2). This financial assistance to home-owners is available on mortgages up to $1 million and includes a second (vacation) home.

The skyrocketing of home-owner subsidies has resulted in a rather anomalous situation. In 1988, households with incomes over $50,000 received 52 per cent of all housing subsidies (Table 2.10). The question might be raised: does any household with an income over $50,000 need a public subsidy? If – as is generally accepted in principle – housing financial assistance should go to those in most need, such favourable treatment of the middle- and upper-income owners may be regarded as a misdirection of housing subsidies, and

*Table 2.10* Estimated distribution of housing subsidies received by various income levels 1988

(subsidies in billions)

| Annual income | Tax expenditures $ bn | % | Housing subsidies $ bn | Total cost $ bn | % |
|---|---|---|---|---|---|
| Under $10,000 | 0.1 | 0.1 | 10.1 | 10.1 | 15.7 |
| $10,000 to $20,000 | 1.1 | 2.2 | 2.7 | 3.8 | 5.9 |
| $20,000 to $30,000 | 3.8 | 7.6 | 1.0 | 4.9 | 7.6 |
| $30,000 to $40,000 | 5.4 | 10.7 | 0.0 | 5.4 | 8.4 |
| $40,000 to $50,000 | 6.6 | 13.0 | 0.0 | 6.6 | 10.2 |
| $50,000 and over | 33.6 | 66.4 | 0.0 | 33.6 | 52.2 |
| Total | 50.6 | 100.0 | 13.8 | 64.4 | 100.0 |

*Source:* Paul A. Leonard, Cushing N. Dolbeare, Edward B. Lazere, *A Place to Call Home: The Crisis in Housing for the Poor*, Washington DC, Center for Budget and Policy Priorities and Low Income Housing Information Service, 1989, p. 33. Data from documents of Joint Tax Committee and the Office of Management and Budget

as one major explanation why more progress has not been made in reducing the affordable housing gap for low-income households.

## Urban decay

Since World War II, the urban community has been undergoing profound changes. With the automobile, suburbia was easily accessible and offered many attractions for better living to those who could afford it. Many traditional sources of employment in the big city left for economic reasons, often leaving a shrunken local tax base. Generous federal subsidies for efficient highway networks and home-ownership accelerated the migration of middle- and upper-income families from the inner city. Large-scale rural to urban migration, particularly from the South to the North, brought a stream of low- and moderate-income families to industrial centres. On the one hand, many large cities were losing population; on the other hand, many cities continued to grow even as certain sectors were in decline. In some cities, massive new capital investment in the central business district led to economic renaissance, while in others stagnation persisted. Midst these myriad changes, in many cities large areas are now engulfed in serious blight.

Urban scholars have long engaged in lively debate about the future of American cities. One school of thought views the decline of the city as an inexorable result of market forces and rapidly changing technology. Another school sees a new type of city emerging from the old – a post-industrial city in which white-collar workers increasingly displace blue-collar workers. Still another view maintains that a vibrant life will always respond to some of the deepest needs of people, and that changes in the urban form are taking place in a more or less normal, evolutionary manner.

Irrespective of the complex economic, social and political forces that have impacted city life, one of the biggest single causes of urban decay is poverty. When households are too poor to pay for decent housing, it is difficult to avoid the ageing of the physical environment and the increase in dilapidated housing, over-crowding, unsanitary conditions and other maladies associated with poverty (McGeary *et al.* 1988).

## METHODS OF SOLVING THE PROBLEM

### Housing the poor

*The housing allowance – an income supplement*

Probably the most important single step in dealing with the housing affordability is to put more money into the hands of the poor so that they can pay the costs of decent housing (Struyk and Bendlick 1981; Freidman and Weinberg 1983). In fact, since 1980 the Reagan Administration favoured this policy in two main HUD programmes: the old so-called Section 8 rent certificate programme and the new housing voucher programme (Kennedy and Finkel 1987).

These programmes make it possible for households to rent minimum standard units in the existing private housing stock. The subsidies reduce housing costs to a fixed percentage – generally 30 per cent – of household income after certain deductions. The households are required to live either in units specifically built for them, or in units that meet federal quality standards. Rents must not exceed Fair Market Rents, as determined by HUD.

Fair Market Rents represent the typical cost of rental housing available in the private market, and are established on an annual basis for all metropolitan areas and non-metropolitan counties. A Fair Market Rent is the dollar amount below which 45 per cent of the standard quality rental housing units in the local market area fall, when listed by rent level. The 45th percentile is drawn from the distribution of rents of all units which are occupied by recent movers, except for public housing and newly built units. The more recent housing voucher programme introduced a special feature: households may choose to live in units costing more than the Fair Market Rent, but only on the condition that they pay the difference above 30 per cent of their income.

The housing allowance concept has many merits. First, it directly reduces the tenant's financial burden. Second, the annual cost per family is less than half that involved in building new housing. Stephen Mayo found that from two to three times as many households could be served per dollar expenditure on housing allowance as could be served by either the public housing programme or by the Section 236 producer subsidy programme for privately owned housing (Mayo *et al.* 1980, 103). In Australia the ratio was also 2:1, while in West Germany it was 4.5:1 (Howenstine 1986, 134–6).

Third, by attaching the housing allowance to the family and not the project, the household has (at least in principle) freedom to choose its own home and its location. Fourth, by choosing its location, minority households have the opportunity (at least in theory) of moving out of inner city ghettos in search of a better, non-segregated living environment. Fifth, it avoids the problems of property management and of possible repossession of housing units associated with producer subsidy systems. Sixth, it avoids inequities inherent in the producer housing subsidy system, which operates as a kind of lottery in making the limited supply of housing units available to a comparatively few poor families. However, until the housing allowance itself becomes an entitlement for all lower income households, in practice it too operates as a kind of a lottery system. Seventh, it channels more income to landlords and thus, following European experience (at least theoretically), promotes better maintenance and repair of the existing stock (Howenstine 1986, 9–11).

On the other hand, the housing allowance system is not without limitations. Its most serious shortcoming is that it does little or nothing – at least in the short-term – to increase the supply of low-cost housing. In localities where there is a severe physical housing shortage, e.g. rapid growth areas or areas experiencing widespread urban decay, the housing allowance is incapable of overcoming the shortage. Second, if in a new housing allowance entitlement programme there is a massive infusion of new money into the hands of renters, this could be very inflationary in a relatively fixed supply situation. Landlords could raise rents beyond what was justified to recoup the costs of making up possible backlogs of maintenance and repair. Third, the system has no way of meeting the needs of large families in communities which lack an adequate stock of large housing units.

Fourth, if the housing allowance is conditional upon the housing unit meeting minimum standards (as it has been, to ensure that people live in 'decent' housing), this may become a deterrent to participation for a large number of eligible families. If receipt of subsidy necessitates a search for, and a move to, a new accommodation, this encounters financial and psychological costs to the tenant, which often outweigh the presumed benefits of living in a standard housing unit. This has been particularly true of the elderly and of ethnic households who attach high value to remaining in familiar, well-established neighbourhoods. In practice, it has often been found difficult to compel a household to spend more income on housing, if it does not really attach that much importance to improving the physical quality of its housing, often in comparatively minor ways. Often, too, there are difficulties in persuading landlords to bring their properties up to standard.

Fifth, there are no built-in cost limits in market situations where housing costs are rising faster than the general cost-of-living index. Sixth, the housing allowance provides little incentive for either households or landlords to economize on costs. Seventh, to ensure that eligibility requirements are being fulfilled, the housing allowance system involves high administration costs in monitoring household incomes and the quality of housing accommodation. Finally, income supplements may do little to overcome discrimination on the part of landlords against non-white households.

Hope – or expectation – that an income supplement in the form of a housing allowance will alone overcome the evils of poverty does not appear to be very realistic. To begin with, based on the experience of The Experimental Housing Allowance Program, only about 50 per cent of eligible families will actually choose to participate. However, even if a generous new housing allowance system were to be adopted and provide an entitlement to all needy households, and even if all qualified families were to participate, the fact remains that, while very important, inadequate housing is only one of the factors contributing to poverty. Not only are there different kinds of poverty but there are also a multitude of factors creating and perpetuating poverty. Moreover, if improvements in housing are not accompanied with a comprehensive anti-poverty programme, including such basic elements as job creation, training and education, better health and nutrition and better home management, then the introduction of a universal housing allowance may have only a limited impact in improving the social fabric.

*Supply measures*

The second method of dealing with the affordable housing problem is to increase the supply of housing (Downs 1983a). During the last ten years there has been a lively controversy over whether there is in fact an actual physical shortage of decent housing. Citing the overall balance between the number of housing units and the number of households in the nation, a national average rental housing vacancy of 7 per cent in 1989, and the comparatively low percentage of sub-standard housing (i.e. 7.1 per cent) in the national housing stock, one school of thought has maintained that, except for special localized situations, there is no nation-wide shortage.

For example, HUD after convening a conference in November 1980 (Weicher *et al.* 1981), concluded:

> There is no need for a nationwide direct subsidy program to stimulate the production of rental housing, in particular, middle-income rental housing. . . . There is no current nationwide shortage in the rental housing market . . .
>
> This is not to suggest that there are no serious problems facing the rental market. The proportion of tenants who are poor and disadvantaged has increased. These poorer tenants may be adversely affected by rising rents which are needed to keep existing multifamily units in operation. For this part of the market, the problem appears to be one of insufficient income, not of market failure in the rental housing market. Further improvements in housing consumption could be obtained more directly by tenant-based subsidies or operating subsidies to existing rental units rather than through new production-oriented programs.
>
> (HUD 1981, i–ii: 4)

Another school of thought maintains that, if one counts the number of people living in sub-standard housing, the number of homeless, the long waiting lists for public housing and the inability of persons to find suitable housing in many metropolitan areas even with Section 8 rent certificates, there is a real physical shortage of affordable housing. For example, the General Accounting Office concludes:

> Millions of Americans cannot afford home ownership and cannot find affordable rental housing. Immediate national attention is necessary if an adequate supply of affordable rental housing is to be available. The Department of Housing and Urban Development is the principal Federal Agency responsible for providing

assistance for rental housing. The Congress and the Administration should take steps to mitigate this nationwide crisis.
(General Accounting Office 1979, 1; Mannheim 1981, 1–13)

The National Low-Income Housing Coalition maintains that until low-income housing needs are met, at least 750,000 units of assisted housing for low-income people should be added to the housing inventory every year (National Low Income Housing Coalition 1986).

Irrespective of whether the shortage of affordable housing is a nationwide or localized phenomenon, there are three major ways to increase the supply of housing: public housing authorities; community-based housing organizations; and subsidized, privately-owned housing.

## (1) Public housing authorities

The public housing programme has been the most important source of permanent affordable housing for low- and moderate-income families; it is especially vital for the elderly, the disabled, large families, and the poorest of the poor. However, for reasons noted on p. 24–5 'Failings of the Public Housing Programme', there is a widely held view that public housing has been a failure and that there should be no more new public housing construction, particularly since the annual cost per family is more than twice that of the housing allowance.

On the other hand, a revitalized public housing programme does have an opportunity to avoid past errors and to make a significant contribution to resolving the affordability housing problem. The Rouse/Maxwell Report offers several recommendations. The most seriously troubled projects should be targeted for immediate remedial action. Modernization of the entire public housing inventory should be completed. Tenant involvement in management should be promoted. Local public housing authorities should be allowed to produce new housing within budgetary constraints (Rouse/Maxwell 1988, 36–8).

The new HUD Secretary, Jack Kemp, has brought quite a revolutionary approach to housing issues in the US. He sees empowerment of the poor as the only successful way to win the War Against Poverty, in fact he exhorts public housing tenants to fight for a future in which they control and eventually own their own housing (Dallas, *Morning News*, 10 December 1989). To reinvigorate the public housing sector, he has established an Office of Resident Initiatives to promote the development of Residential Management Corporations with responsibility for project management functions, such as maintenance, security and rent collection (Caprara 1989; Scanlon 1990). By January 1991, over 100 resident management entities were being developed, with an aim of having 250 in training by 1992. The overall thrust of the programme, however, has been a redisposition of existing programmes toward the most needy, and not an increase in total production of housing for low- and moderate-income families.

Stegman and Sumka have drawn attention to an important, but much overlooked, consideration: 70 per cent of local housing authorities, one-half of all projects, and one quarter of all public housing units are in non-metropolitan areas, that is, cities with less than 50,000 population. These projects have not been overwhelmed by the negative forces which have turned some inner city projects into problem ghettos. The development costs of non-metropolitan housing have been 30 per cent less than the national average, operating costs 41 per cent less, and the federal cost of amortization 56 per cent less (Stegman and Sumka 1976, 258). Where there are clearly defined supply shortages, new public housing construction programmes in non-metropolitan urban areas remain a cost-effective alternative in providing affordable housing.

Rachel Bratt makes some additional recommendations to avoid the errors of the past. Construction costs should be financed by 'up-front' capital grants rather than by interest subsidies over the lifetime of the project; advance provisions should be made to supplement operating budgets and to meet modernization costs before financial crises arise; the programme should include appropriate social services for the tenant population; and all new public housing design should be consistent with the housing design of the surrounding neighbourhood (Bratt 1989, 321–2).

(2) Community-based housing organizations and public–private partnerships

One fortuitous result of the Reagan Administration's retrenchment in federal housing policy has been the increased stimulus for developing community-based housing organizations (CBHOs), i.e. organizations in which members of a local group, such as a church, civic association or labour union, or a group of tenants, join together to produce, rehabilitate, manage or own housing. They are generally non-profit and their primary orientation is toward low- and moderate-income families in their neighbourhood. In fact, for several decades

HUD has been involved in providing financial and technical assistance to CBHOs under various national housing acts. Some, such as the non-profit projects for the elderly providing capital loans at 3 per cent (Vanhorenbeck, 1989), have been very successful, while others have had a rather checkered performance.

Then in the 1980s, a new generation of over 4,000 community organizations, made up of community development organizations, non-profit developers, religious institutions, labour unions and other CBHOs have sprung up to undertake long-term housing projects for low- and moderate-income residents that were considered too risky or too small by financial institutions or other for-profit developers (HUD 1987a; HUD 1989b). Needed capital has often been provided by special Housing Trusts set up by State and local governments (Brooks 1988; Peirce and Steinback 1987). In addition the Neighborhood Reinvestment Corporation established by Congress in 1978 has inaugurated a modest Neighborhood Housing Services Program in 137 cities, providing loans and public improvements to CBHOs for housing rehabilitation (Neighborhood Reinvestment Corporation 1989). The Local Initiatives Support Corporation, with initial capital support from the Ford Foundation, has provided technical and financial resources to over 500 community development corporations for housing and other economic development projects in deteriorating communities (Local Initiatives Support Corporation 1989). The Enterprise Foundation, founded by James Rouse, offers technical, management and financial assistance to over seventy CBHOs (Enterprise Foundation 1989). The Housing Assistance Council has provided technical assistance, training and seed money to rural CBHOs since 1971 (Housing Assistance Council 1989). Habitat for Community, to which former President Jimmy Carter has been deeply committed, has organized volunteers to build over 3,000 homes for the poor throughout the country, and has developed a network of overseas projects (Habitat for Humanity 1990). The Boston Housing Partnership is a good example of how three important parties – the banking and business community; local and state public officials; and non-profit community development corporations – can cooperate in rehabilitating existing low-income housing (Bratt 1989, 290–319).

For CBHOs to have a significant impact on the national housing supply situation, however, new resources and leadership are required. Thus the Rouse/Maxwell Report recommends that the federal government establish a new national corporation to promote and support CBHOs, with an emphasis on Benevolent Lending, i.e. encouraging individuals, corporations, churches, foundations and others to lend funds at rates as low as 3 per cent to finance low-income housing (Rouse/Maxwell 1988, 26–8).

Bratt lists four main ingredients for CBHO success. First, adequate financial resources are crucial, including specifically (a) seed money for organizational expenses, (b) first-step funding for project initiation, (c) construction and debt financing for project implementation, and (d) consumer subsidies to reduce costs to households. A second essential is the CBHO's ability to gain control of land and buildings at affordable prices. A third ingredient is technical assistance. Finally, there must be a communication and dissemination network, including on-going research and evaluation, to share and profit from experience (Bratt 1989, 276–82; Stegman *et al.* 1987).

(3) Tapping the private sector through tax incentives and density bonuses

As noted previously, a large segment of the housing stock occupied by low- and moderate-income families is privately-owned but publicly-subsidized through various federal programmes. Tax concessions have been an effective tool to induce individuals and corporations to invest in low-income housing. Through accelerated methods of depreciation, owners can generate huge paper losses on their rental properties which can be offset against other ordinary income in calculating income taxes. By selling shares in limited rental property partnerships through syndicates, large developers can tap capital markets for persons seeking such 'tax shelters' (Verdier 1977).

To ensure that the housing is occupied by low- and moderate-income families, partnerships must accept use restrictions, i.e. the requirement that low-income household occupancy be maintained for a specific period, ranging under various programmes from 20 to 40 years. At the end of the use period, the developer is free to convert the units to up-market renters or to condominium ownership. As periods of use restriction begin to draw near for many projects, difficult issues arise: (1) will rents be raised? (2) will existing tenants be displaced? (3) will this sector of rental housing be converted to ownership status? The status of this part of the subsidized housing stock has been very much debated (Low Income Housing Information Service 1990).

The Tax Reform Act of 1986 placed restrictions on the above tax shelters, and introduced a new tax programme for developers who make a proportion of units available for low-income households. If 40 per cent of the rental units are made available to households with incomes below 60 per cent of area median income, (or 20 per cent of the units to households with incomes below 50 per cent) the developer (or investor) may take a yearly tax credit of 4 or 9 per cent of the cost of the low-income units (depending on whether other federal subsidies, such as tax-exempt bond financing, are received) for ten years. The programme achieves a measure of social and economic integration by mixing subsidized low-income tenants with market-rate tenants (Taylor 1987; Gravelle 1987).

Another tax incentive has been federal permission to state and local governments to issue bonds which are exempt from federal income taxes, the proceeds of which are used to finance below-market rate mortgages. For example, in 1985, tax exempt bonds financed an estimated 300,000 rental units, at least 60,000 of which were set aside for families with incomes at 80 per cent of area median income or less (Rouse/Maxwell 1988, 40).

A further device which has been increasingly used by local governments for expanding the supply of affordable housing is the granting of density bonuses; that is, giving developers the right to build more housing units on a given land parcel than had been originally provided in the zoning ordnance, on the condition that they reserve a certain proportion, e.g. 20 to 40 per cent, of the units for low-income tenants. The Bridge Housing Corporation has been a highly successful exponent of this system in the San Francisco Bay area (Bridge Housing Corporation 1989).

## (4) Reducing construction costs

One of the long-standing battles in making housing more affordable is the reduction of construction costs (Howenstine 1983). In 1982, HUD launched a Joint Venture for Affordable Housing, a nation-wide demonstration programme searching for ways to achieve cost savings by relaxing normal developmental regulations and simplifying approval procedures. Over 100 demonstration projects were completed, with an average cost saving per unit of $8,573 (HUD 1987c).

In March 1990, HUD Secretary Jack Kemp appointed a blue-ribbon Advisory Commission on Regulatory Barriers to Affordable Housing to make recommendations for the elimination or revision of excessive or unnecessary federal, state and local regulations. The review will include zoning, impact fees, subdivision ordinances, codes and standards, rent control, permits and processing, and federal and local environmental regulations. The Commission will report its findings within one year.

There is a widespread consensus that unremitting efforts are required to lower land, construction and financial costs of housing.

### Homelessness

Community response to the homeless has on the whole been positive. A 1988 HUD survey showed that the number of homeless shelters in the US had almost tripled since 1984, while the money being spent on shelters had increased five-fold. Shelter efforts have been a grassroots phenomenon strongly characterized by volunteerism and joint public–private partnerships. Ninety per cent of the shelters are operated by private, non-profit organizations, aided by many volunteers, while two-thirds of the financial support comes from local, state and federal governments (HUD 1989b, i; HUD 1989c).

In 1987, Congress passed the McKinney Homeless Assistance Act which provided $1 billion over a two year period for a series of programmes to increase shelter construction and rehabilitation assistance, to provide transitional housing assistance to help homeless persons move from dependency to self-sufficiency, and to stimulate the development of single room occupancy hotels (Miles 1988). These steps, while sound, are generally recognized as being stop-gap measures. A really serious attack on the problem is a tall order; as a minimum, it would require permanent housing plus comprehensive job and services support programmes.

One modest policy of strategic importance that many state and local governments have adopted in recent years is the provision of emergency financial aid to cover unpaid rents in order to avoid eviction, 'eviction prevention' as it is called. Low-income families are under constant threat of homelessness, e.g., when the breadwinner is thrown out of work or some member of the family has a serious illness. Everyone benefits by helping vulnerable families retain their homes, particularly the local government for whom a displaced family becomes very costly to support on welfare.

### Neighbourhood revitalization

Combatting urban decay has been high on the agenda of US political action for a long time, and indeed some good results have been achieved. But the record is patchy. In the early post-war period, the federal Urban Renewal Program was the centrepiece of urban policy. The central concept was that replacing slums with decent housing was the way to improve the quality of life and reshape society. However, it floundered on the inhumanity of bulldozing slums, displacing the poor and redeveloping areas for middle-class and commercial occupancy (Anderson 1964; Wilson 1966).

In 1964, President Johnson launched his War on Poverty, which, *inter alia*, challenged communities to create long-range plans for attacking poverty and provided funds for establishing Community Action Agencies. However, serious conflict soon arose between local citizen groups representing the poor and the city mayors who controlled the political machines and coordinated financial resources.

Then in 1966, Johnson embarked on a second major effort, the Model Cities Program. In contrast to the Urban Renewal Program which focused on rebuilding the physical environment, the Model Cities Program visualized a massive, five-year attack on the social environment of the most troubled areas within major cities. Although a great amount of professional wisdom went into its making, in retrospect the Model Cities concept never received a fair chance. To begin with, the proposed redirection of federal aid systems to the most needy was very ambitious (and perhaps an unrealistic goal, considering the vested interests in the existing federal aid structure). Moreover, in practice not only were the promised increased resources eroded by the escalating financial demands of the Vietnam War, but the projected concentration of social and economic programmes in the distressed areas of a comparatively limited number of cities was diluted by a great increase in the number of participating cities as a result of political pressures. Finally, HUD's ability to coordinate and integrate around fifty federal programmes administered by seven Cabinet Departments proved ineffectual (Frieden and Marshall 1975; Sundquist and Davis 1969).

In the 1970s, the Model Cities Program was phased out, as the revenue-sharing philosophy of President Nixon spawned the new Community Development Block Grant programme. The federal revenue-sharing policy provided income to state and local governments, which was virtually free of federal requirements, and which could be used for tax relief or to fund any of the normal government activities. Increasingly the key question was raised: can top-down social reform really regenerate the social fabric? Or must it be bottom-up to be effective?

The most comprehensive recent study of urban decay and decline is that of the Brookings Institution, which recommends two types of action: changes in institutional structures; and measures designed for individual households. As regards institutions, they recommend (1) removing anti-city biases from federal policies, specifically funding for highways, transit, sewer and water systems; locating more federal jobs in the target areas; and tax benefits for housing; (2) coping with regional biases in federal policies; (3) helping cities adapt to smaller population and fewer jobs; (4) providing more effective forms of federal aid to cities; and (5) overcoming fragmented urban jurisdictions through the creation of metropolitan-wide governments.

A second set of measures is recommended to empower individual households threatened or injured by urban decline. The most comprehensive form of empowerment would be a combination of jobs and guaranteed annual incomes for unemployed and destitute households. More limited forms would consist of voucher systems for particular purposes, such as employment, housing, public transit, education and health. To attack the problems of concentrated poverty resulting from socio-economic and racial segregation, which is supported by institutional arrangements that benefit a majority of metropolitan-area citizens, Bradbury and Downs believe that there must be some movement of the inner city poor into more dispersed locations throughout metropolitan areas, or at least into moderate-size clusters in most parts of each area (Bradbury *et al.* 1982, 14–7).

Two types of neighbourhood revitalization have been developed to cope with urban decay: 'incumbent upgrading' and 'gentrification' (Bradbury *et al.* 1982, 212). Both types start with improvements in physical structures, more attractive local environments, declining crime rates, rising property values, and increased confidence on the part of investors and property owners. But they differ sharply regarding the occupants of the rehabilitated structures. Incumbent upgrading is a process through which existing low- and moderate-income families rehabilitate deteriorated areas for themselves. One of the major vehicles for revitalization is of course the CBHO. From the standpoint of living conditions of the poor, this is certainly the preferred route.

On the other hand, gentrification is the process in which more affluent newcomers buy and refurbish homes in run-down neighbourhoods. Local governments often welcome this kind of revitalization. It brings back into the inner city middle-income people with talents, skills and energies which raise the tempo of local economic activity. Moreover, the accompanying increase in the community's tax base enhances the city's financial capacity to make needed improvements in its infrastructure and services. Professor Gale sees this type of revitalization as the logical forerunner of the post-industrial city in which white-collar employment occupies a growing share of the metropolitan labour force (Gale 1984).

**First-time home-buyers' fading dream**

Although over 2 million young families have been backed – at least temporarily – in realizing their dream of home-ownership, their plight is of quite another dimension than the financial burden on the poor. To begin with, home-ownership already has preferred treatment. There is no tax on the imputed income from capital invested in the home (as there is in many European countries), although the income from practically all other forms of invested capital is taxed. Also, home-owners are entitled to full tax deductibility of mortgage interest and property taxes, while only a quarter of the poor receive some form of housing financial assistance, which is not an entitlement. In addition, two programmes provide extra home-ownership subsidies for first-time, moderate-income home-owners. States and localities have a limited power to issue federally tax-exempt bonds, the proceeds of which may be used to finance mortgages at lower than market interest rates (Appendix 1). There is also a small new Nehemiah programme ($24 million in 1990) which extends a subsidy to first-time moderate-income home-buyers through a second mortgage up to $15,000, on which no interest accrues, and on which repayments are deferred.

Even though many young people today may not be in a financial position to embark on home-ownership, most are able to afford rental housing. Furthermore, the frustration suffered by aspiring home-owners as a result of rising home prices has happened before; in 1946/7, 1965, and 1977 (Weicher 1980, 92–109). Moreover, while in the present crunch many young people may not be able to become home-owners immediately, it should be pointed out that their parents normally had to save up money for the down deposit over a five to six year period. It may be noted, too, that the size and amenities of contemporary homes, which many of the present generation have grown to expect, are at a much higher level and much higher cost than that with which their parents started out a generation ago. Many young households might be satisfied to start off with a smaller, stripped-down home.

Nevertheless, while it may not be feasible to provide an additional direct subsidy, e.g., an up-front, one-time capital grant as some European countries do, there are some practical alternatives. First, certain changes in the traditional mortgage instrument offer lower initial monthly costs. The shared appreciation mortgage (SAM) and the equity-adjusted mortgage (EAM) in effect allow households to implicitly (through reduced payments) and explicitly (through actual annual loans) borrow against the expected appreciation of their properties. The graduated payment mortgage (GPM) sets up a payment schedule that permits lower payments in the early years and higher payments in later years, when presumably the household's income will have increased. Adjusted-rate mortgages (ARMs) lower initial costs by offering two to three percentage points below fixed rate instruments (Rosen 1984, 127–41).

Second, there are also ways to deal with the down-payment problem. For example, following the Rouse/Maxwell Report recommendation, President Bush proposed an amendment to the Individual Retirement Account law, which will permit use of the retirement fund savings for a first-time home purchase without tax or penalty. Based on the highly successful German *Bausparkassen* concept, Goetz has suggested the setting up of Individual Housing Accounts, which could be accumulated through annual tax-free savings of, say, $2,500, specifically for first-time purchase of a home (Goetze 1983, 122; Bourdon 1988). The Rouse/Maxwell Report has also recommended reducing down-payments by permitting Federal Housing Administration insured buyers to secure financing up to 97 per cent of the first $50,000 of a home's value, and 95 per cent of the value in excess of $50,000, as well as by raising the mortgage limits above the present FHA maximum of $101,250 (Rouse/Maxwell 1988, 45–9).

By way of a concluding perspective, it may be noted that expansion in home-ownership has certain indirect benefits for low-income renters. Any movement of households from the rental to the ownership sector immediately frees up space and thus relieves some price pressure on the rental market. Moreover, in so far as first-time buyers purchase new construction, this adds to the total supply of housing and through the 'trickle down' process tends to relieve pressures throughout the entire housing market.

## THE DISMANTLEMENT OF NATIONAL HOUSING POLICY: THE REAGAN ERA

President Reagan came into office in 1981 with the belief that, as succinctly formulated by the President's Commission on Housing,

> the genius of the market economy, freed of the distortions forced by government housing policies and regulations that swung erratically from loving to hostile, can provide for housing far better than Federal programs.
>
> (President's Commission on Housing 1982, xvii)

He viewed his election as a popular mandate (to put it simply, albeit with a degree of exaggeration) to get the Federal Government out of the housing business. While complete success was not achieved, the eight years, 1981–9, witnessed significant disengagement (Struyk *et al.* 1983; Peterson *et al.* 1986a and 1986b; Hartman 1986; Liner 1989).

Probably the first major target was the public housing programme. After its 1983 budget request, the Administration did not request any more funds for new conventional public housing. The relatively few new units which were approved by Congress during the succeeding years were constructed over the Administration's continuing objections. The number of annual additions to the public housing stock fell from around 44,000 in 1979 to 5,200 in 1989 (Table 2.9). Then in 1985, inspired by Prime Minister Margaret Thatcher's policy in the United Kingdom, HUD launched a demonstration programme to privatize public housing (Rohe and Stegman 1990; Howenstine 1985).

Another important part of the HUD structure failing to find favour with President Reagan was the Office of Neighborhoods, Voluntary Associations and Consumer Protection, an office which promoted action on behalf of tenants. This office was dropped in 1981.

A third HUD policy believed to be incompatible with the American free enterprise philosophy was urban planning. In its formative years, America has never managed to develop the tradition and practice of town and country planning, such as Western European countries evolved through the centuries, giving European cities their distinctive quality. Nevertheless, feeble, piece-meal efforts began to appear in the 1930s. The principal recent HUD effort in this field was the Comprehensive Planning Assistance Program (Section 701), which over its lifetime provided over $1 billion in grants to states and local and regional organizations to promote orderly and efficient urban growth and development. Already on the defensive after the Community Block Grant Program was adopted in 1974, this programme was abolished in 1982.

Two other important elements in urban planning policy were also phased out in the 1980s: revenue sharing; and the Office of Management and Budget planning review procedure (Circular A95). Revenue sharing had been the major financial source for the planning activities of thousands of small towns in America. The A95 review procedure had helped to ensure that federal construction grants to state and local governments were not in conflict with regional development plans, particularly transportation networks.

Another aspect of HUD's policy came from a Congressional initiative in 1977, calling on the President to prepare a biennial Report on National Urban Policy. The purposes were to examine urban trends and problems and to present a comprehensive national urban policy to deal with these problems. The Reagan Administration continued this effort. However, as the Administration believed that urban policy was not an area in which the federal government should intervene directly, the Report was confined to a review of trends and problems, and did not put forward a national policy to solve the problems (HUD 1988, 98–9).

A fifth important HUD policy destined for overhaul was the Minimum Property Standards for One and Two Family Dwellings. These standards, which had had their last major revision in 1974, are one of the basic requirements for Federal Housing Administration loans. In 1982, a new set of standards and regulations was issued, removing those pertaining to marketability and liveability (i.e. saleability and design) of housing units, and leaving only those relating to life safety and health.

A sixth notable retrenchment came in regard to housing research and the housing data base. With disengagement being the banner, it was believed less important to have a vigorous, inquiring research arm and a streamlined information system. From 1979 to 1989, the HUD research budget fell from $51 million to $17 million, while the research staff declined from 252 to 134. As the thrust of the research programme was blunted,[2] the centre of the nation's urban research was increasingly dispersed to other non-executive controlled government offices, such as the Congressional Budget Office and the Congressional Research Service, to non-profit research centres in Washington DC such as the Urban Institute, and to academic research centres throughout the country, such as the Joint Center for Urban Studies at Harvard University and the Massachusetts Institute of Technology. Meanwhile the last edition of the *HUD Statistical Yearbook* to appear was in 1979.

Disengagement was even more pronounced with regard to HUD's participation in the international research arena. During most of the postwar period, HUD and its predecessor organization, the Housing and Home Finance Agency, had an important substantive involvement in the two principal international forums, the Committee on Housing, Building and Planning of the United Nations Economic Commission for Europe based in Geneva, and the Group on Urban Affairs of the Organization for Economic Cooperation and Development based in Paris. During the 1980s, however, as a result of reduced budgets and internal bureaucratic stresses, HUD's research participation gradually dwindled until in 1989, with the liquidation of the Office of International Affairs, it became non-existent. Meanwhile the Foreign Information Retrieval System in the Office of International Affairs was disbanded in the early 1980s.

Overall, President Reagan accomplished a great deal of cutting within HUD. He trimmed HUD's budget authority (i.e. the long-term federal financial commitment involved in implementing housing programmes adopted by Congress) to around $8.8 billion in 1989 from around $41 billion (in 1989 dollars) in 1980 (Pedone 1988, 42). He lowered the cost of federal housing subsidies significantly by raising the percentage of income that low-income families had to pay for rent from 25 per cent to 30 per cent. He cut total HUD staff from an average of 15,500 during 1977–9 to an average of 12,500 during 1987–9.[3] He reduced the number of assisted housing units newly subsidized each year through additional Section 8 rent certificates from an average of around 231,000 during the three year period, 1977–80, to an average of around 92,000 per year in the three year period, 1987–9 (Table 2.2).

Nevertheless, perhaps surprisingly, financial outlays for HUD and Farmers Home Administration subsidized housing programmes rose steadily from $6.9 billion in 1979 to $16.7 billion in 1989 (Tables 2.3 and 2.4). This was due to the steady increase in the number of households receiving assistance, to inflation, and to the fact that much of the money appropriated in the 1970s was not actually spent until the 1980s.

In retrospect, it may be observed that while President Reagan did not promote affordable housing through a vigorous national housing policy, he did 'pledge to foster good housing for all Americans through sound economic policies' (President's Commission on Housing 1982, vx).

## REBUILDING NATIONAL HOUSING POLICY

During the last three years, there has been a fundamental re-examination of the Reagan disengagement policy. First, the national Housing Task Force, created by Congress in September 1987 on the initiative of Senators Alan Granston and Alphone d'Amato (the Chairman and the Ranking Member of the Senate Subcommittee on Housing and Urban Affairs) to undertake a National Housing Policy Review, issued a path-breaking Report in March 1988 (Rouse/ Maxwell Report 1988). The Report drew on a series of twenty outstanding, especially commissioned technical monographs by the country's leading housing analysts (Schussheim, 1988). The National Low Income Housing Preservation Commission, under the Co-Chairs, Carla Hills and Henry S. Reuss, made its Report to Congress in June 1988 (National Low Income Housing Preservation Commission 1988). The Congressional Budget Office issued a comprehensive study in December 1988 (Pedone 1988).

A National Affordable Housing Act, sponsored by Senators Cranston and d'Amoto, along with thirty other senators, was introduced in April 1989 (Schussheim 1989). The General Accounting Office convened a high-level Housing Conference in August 1989 (General Accounting Office 1989). In November 1989, President Bush presented his HOPE Initiative, i.e. Homeownership and Opportunities for People Everywhere (HUD 1990a). The National Low-Income Housing Conference was convened in February 1990 (National Low-Income Housing Coalition 1990), and a month later the National Housing Conference met (National Housing Conference 1990). Following up on the Rouse/Maxwell National Housing Policy Review – with the support of the Ford Foundation – the Urban Institute launched an outstanding series of nine policy seminars in 1988–9; the findings were published in 1990 (Turner and Reed 1990). Independently, a number of challenging studies by housing specialists have been published: they aim to rebuild a comprehensive national housing policy responsive to the critical needs of low- and moderate-income families.[4]

The central theme running through most of these activities is summed up by the Rouse/Maxwell Report: 'What is missing is adequate participation by the federal partner. . . . The first step is commitment. The Federal Government must reaffirm its role as a leader in finding solutions to the country's housing problems' (Rouse/Maxwell 9–10).

Squarely facing the 1949 Housing Act goal, the National Housing Task Force affirms: 'We must fire up the energy and fulfill the legitimate expectations of decent housing for all our people . . . by the year 2000 – in 12 years' (Rouse/Maxwell, 3).

There is a general consensus among housing analysts concerning the broad principles of national housing policy for the 1990s, though there is much disagreement on how they should be implemented.

## The consumer subsidy approach

The first most effective step in attacking the affordable housing problem is to expand the housing allowance programme. The central issues are costs and timing. Should such a programme be increased incrementally, e.g., by adding a certain number of new low-income families each year, say 200,000 to 400,000 or more, until the year 2000 when all qualified persons would have an entitlement, as proposed by the Rouse/Maxwell Report? Or should it be converted to a universal entitlement programme in one massive change within the near future?[5]

One major factor affecting the cost of such a programme is the participation rate. In the HUD Experimental Housing Allowance Program demonstration during the 1970s, roughly 50 per cent of the qualified persons opted to participate. In any further entitlement programme, this rate may well be higher and more in line with European experience of around 75 per cent (Howenstine 1986, 62–3) for several reasons. Benefits may be higher; benefits will more likely be perceived as permanent (i.e. not for only a limited period); and since more households will already be occupying standard housing, there will be less need to correct physical housing deficiencies as a condition for receipt of allowance or to consider moving to a minimum standard housing unit.

In 1989, HUD spent almost $10 billion on housing allowance types of programmes (Table 2.4). There seems to be agreement that an entitlement programme serving all eligible households (assuming around a 50 per cent participation rate) would cost an additional $17 to $23 billion annually. If, however, assistance were to be limited to households with only the most severe needs, the programme still would require an additional $13 billion (Turner and Reed 1990, 34; Pedone 1988, 104–9).

The financing of such a programme is, of course, a question of political acceptability. It is not a matter of objective economic assessment of the nation's resource capacity. Nor is it viewed as a matter of the moral responsibility that middle- and upper-income families should have in sharing some of their well-being with poor families. The political reality is that few, if any, politicians believe they can be elected if, in advance, they proposed a significant increase in taxes to meet urgent social needs. Some day this mood may change. Meanwhile, however, it is generally agreed that the means for financing new housing programmes must be found in reallocating current ventures among existing programmes.

One major part of the federal budget where financial resources are being freed up is military expenditure, the so-called 'Peace dividend'. Estimates concerning the size of the peace dividend vary widely, and there is a multitude of worthy purposes being nominated for selection. How soon the dividend will materialize, and whether housing will be favoured in any distribution, is hard to predict, but another source of funds is being increasingly discussed: a reasonable reduction in the size of the subsidies which home-owners receive in income tax deductibility of mortgage interest and property taxes. Currently, there is considerable informal and unofficial support for this idea around Government, e.g. in the Department of the Treasury, the Office of Management and Budget, the Government Accounting Office and the Congressional Budget Office, and for a long time it has had the general support of housing analysts. In 1969, when HUD secretary George Romney proposed that the deductions be disallowed, total home-ownership deductions for mortgage interest and property taxes were only $2.9 billion![6] With tax subsidies in 1991 expected to be almost $60 billion (Table 2.3),[7] the cost of an immediate $17 to $23 billion housing allowance entitlement programme could – at least in principle – be financed by a 30 to 40 per cent reduction in existing home-ownership tax subsidies.

If such a policy were to be adopted, there are two broad options for its implementation (Downs 1983, 144–5; Rosen 1984, 167; Struyk *et al.* 1988, 9; Muth 1981, 153–4). The first and most equitable method would be to convert the tax deductibility into a tax credit. The tax credit system provides every taxpayer with the same amount of tax savings for each dollar paid for mortgage interest and property taxes regardless of the taxpayer's marginal tax rate. Existing tax deductibility, on the other hand, provides much greater savings to wealthy taxpayers who are in higher marginal tax brackets.

If switching the tax deductibility to a tax credit is not politically acceptable, the second option, following the practice in Austria, Federal Republic of Germany, Finland, France and the United Kingdom (Howenstine 1983, 44), would be to place restrictions or a cap on allowable deductions. At least three countries – Australia, Canada and New Zealand – have no tax deductibility of mortgage interest or property taxes.

About two-thirds of the current tax benefits go to households with annual incomes placing them in the top 11 per cent of all tax payers (above $50,000), and nearly 90 per cent go to households above median income (Miles 1990, 9). Deleting the second house and lowering the cap below $1 million sufficient to yield $23 billion would seem to be a feasible proposition. It is hard to make the case that middle- and high-income households need these subsidies more than low-income households.[8]

An additional reason for reconsidering the wisdom of home-owner tax subsidies emerges from a recent Urban Institute study. Demographic projections indicate that the number of elderly and non-elderly individuals living alone will grow very rapidly over the next four decades, from about 28 per cent of households in 1985 to around 35 per cent by 2010 (Struyk *et al.* 1988, 280; Burns and Grebler 1986; Hughes and Sternlieb 1987). These trends imply an increased demand for smaller units, and a sharp reduction for large single-family houses. From a national policy point of view, therefore, there may be some merit in slowing down the rate of increase in over-consumption of housing space by affluent home-owners in outlying suburbia. This would facilitate a better adjustment to demographic and post-industrial settlement trends and help avoid the possibility of an excess of large single family houses in the twenty-first century (Struyk *et al.* 1988, 5).

There is no great promise that legislative action on this matter will take place in the immediate future. Few politicians are ready to support such a policy in the face of powerful real estate lobbies that can appeal to the 64 per cent of the households which own their own homes – even though only about 53 per cent of home-owners actually took mortgage interest deductions in 1986.[9] Nevertheless, the political situation could quite easily change (Saunders 1989, 119–21).

Another consideration justifying a reduction in home-ownership subsidies is that it could be of some assistance in making housing prices generally more affordable. The availability of virtually unlimited tax subsidies has had an important escalator effect on housing prices, as people have moved up-market in pursuit of more housing space and greater capital gains. Housing analysts have estimated that housing prices are about 15 per cent higher than they would be if deductibility did not exist (Miles 1990, 12).

### Expanding the supply of affordable housing

The other main pillar in national housing policy is preserving and expanding the supply of low-cost housing. First in order of business is to preserve the existing supply of subsidized housing. A considerable part of the public housing stock is run down and in need of modernization. According to the best recent estimates, the cost of bringing the public housing inventory up to standard will range from $13 to $22 billion, depending on the level of rehabilitation (HUD 1990b, ES–5).

The privately-owned, federally-subsidized low-income housing stock is also in jeopardy, first, because of early withdrawal of housing from low-income programmes through the provision for advance repayment of assisted mortgages; second, because of expirations in use restrictions attached to projects; and third, because of a backlog of needed rehabilitation. To keep this low-income housing stock intact, it is essential that new federal financial arrangements be worked out that will provide a reasonable accommodation between the legitimate rights of owners and the occupancy interests of the tenants. In 1988, the National Low Income Housing Preservation Commission estimated that it would cost $11.3 billion in new funding to preserve for fifteen years 473,000 of the 523,000 units believed to be in danger of loss, to protect the 50,000 displaced households, and to cover $6.4 billion for continuing subsidies previously granted to these properties (National Low Income Housing Preservation Commission 1988, 10).

The other large sector of housing occupied by low-income families is privately-owned but not subsidized. Approximately three out of every four poor families who receive no federal housing assistance live in privately-owned housing and apartments. In 1983, 80 per cent of all renter families with incomes below $10,000 lived in buildings with fewer than 20 units; 60 per cent lived in buildings with 4 units or less, or in single family houses (Rouse/Maxwell 1988, 34). As this housing stock ages, more attention needs to be given to maintaining it in good condition, so that it is not permanently lost. This is of fundamental importance in public policy, since the cost of existing housing is practically always lower, generally far lower, than the cost of new construction. Many landlords will not be able to obtain from low rents the financial resources required to keep these properties in a good state of repair. Recognizing this problem, the Rouse/Maxwell Report proposed that federal, state and local governments develop programmes which will make grants and loans available to cover part of renovation costs (Rouse/Maxwell 1988, 34–5). Also an expanded housing allowance programme could be an important factor in enabling tenants to pay higher rents; this will put landlords in a stronger financial position to modernize and rehabilitate ageing properties.

Although much success may be achieved in maintaining the existing supply of low-income publicly and privately-owned housing, that may not be enough. New construction may often be needed to overcome critical local shortages. What, therefore, are the most effective ways to add new units to the affordable housing stock, taking account of economic, social and political considerations?

First, there is the private sector. As a general rule, the private sector cannot be expected to provide housing to low-income families at rents that will cover all costs. Nevertheless, with a selective use of tax incentives and density bonuses, private developers, banks and corporations can continue to make an important contribution to solving the affordability problem. Private sector housing has the virtue of avoiding the complexities of public ownership and management, and often of being able to achieve a quiet mix of low-income families with market-rate tenants.

Second, community based housing organizations have demonstrated an expert capacity not only to rehabilitate to existing rental housing but also to construct new housing. The Rouse/Maxwell Report and HUD Secretary Kemp see the CBHOs playing an important role in the 1990s.

Third, there is great merit in promoting new construction through individual and cooperative ownership programmes among moderate-income families based on shallow subsidy systems, i.e. low level subsidies. Moderate-income families are able to shoulder a larger share of new construction costs than low-income families – thus lowering the final cost to the taxpayer. Furthermore, they are more likely than low-income families to maintain the new housing stock in a good state of repair. Any movement of moderate-income families into new housing then frees up more space in the existing housing stock for low-income families. Secretary Kemp has in fact gone further. With a goal of creating 1 million new first-time and lower-income home-owners by 1992, HUD has set up a multi-phased home-ownership programme to privatize public housing and other defaulted properties acquired in various Departmental housing programmes (HUD 1991a; HUD 1991b).

Fourth, new units can be added to the low-income housing stock through public housing authorities. As noted before, there is much opposition to a revival of the public housing programme along traditional lines. If public housing projects are built outside the inner city, there is usually strong neighbourhood resistance; if they are built in the inner city, they often run the risk of being engulfed in urban decay and of compounding the social and economic problems associated with the high concentration of poor families. It would seem appropriate, therefore, to limit the scope of any new public housing programme, particularly in a period of financial austerity. As a minimum, there seems to be general agreement that any new public housing should not only be scattered and harmonized with existing neighbourhood design, rather than in big projects, but should also be set up with maximum tenant participation and possibly tenant management and control.

In many cities there is an urgent need to provide more large housing units for large families who cannot find suitable space in the existing housing stock; the numbers involved, however, are relatively small. New housing is also widely needed for the rapidly ageing population, though in some areas it may actually have been overbuilt. In some metropolitan areas facing a critical shortage, the expansion of the public housing stock might be better pursued by leasing or buying existing privately-owned housing than by undertaking new construction. The cost basis of existing housing will almost always be substantially lower than the cost of new housing. In non-metropolitan areas where physical shortages of low-income housing clearly exist, new public construction may also play an important role.

## THE NATIONAL AFFORDABLE HOUSING ACT OF 1990

### The structure of new programmes

A concatenation of forces stirring in the late 1980s (see Section VI) came to fruition on 28 November 1990, when the Granston-Gonzalez National Affordable Housing Act (NAHA) became law. The most important housing legislation in two decades, it signals a market change of direction in national policy. First of all, it reaffirms the national goal that every American family be able to afford a decent home in a suitable environment. Then building on existing programmes, it assembles a wide range of policy initiatives, the most important of which are: the HOME Investment Partnership; the Homeownership and Opportunity for People Everywhere (HOPE) programme; the preservation of existing affordable housing threatened by mortgage pre-payment; the National Homeownership Trust; and housing for people with special needs.

The NAHA centrepiece (promoted by congressional leadership both Democrat and Republican) is the HOME Investment Partnership, a new block grant programme in which funds will be allocated to state and local governments on a formula basis according to federal money (25, 33 and 50 per cent, depending on the programme), with 60 per cent of the funds going to localities and 40 per cent to states. For fiscal year (FY) 1991, $1 billion has been authorized, and $2.1 billion for FY 1992. Funds can be used to acquire, rehabilitate and construct affordable rental and ownership housing. Ten per cent of FY 1991 funds and 15 per cent of FY 1992 funds had to be set aside for new construction or substantial rehabilitation of rental housing. Rental housing will be considered affordable under the HOME programme only if rents do not exceed Fair Market Rents or 30 per cent of 65 per cent of area median income, whichever is lower. State and local governments must prepare a comprehensive affordable housing strategy, setting forth five year projections of housing needs and an action plan for meeting these needs. At least 15 per cent of the funds must be set aside for non-profit, community-based housing organization.

NAHA's second major thrust (promoted by President Bush and Secretary Kemp) is the Homeownership and Opportunity for People Everywhere (HOPE) programme. Believing that home-ownership is superior to rental housing in cost effectiveness and resident satisfaction, it seeks to empower the poor through resident management and ownership of public housing. The Act authorizes $448 million in grants over two years to promote private ownership of single-family public and Indian housing. The subsidy takes two forms: (a) planning and technical assistance grants, limited to $200,000 per project, to assess project validity and prepare residents for home-ownership; and (b) implementation grants for rehabilitation as well as for counselling and training, economic development activities, capital reserves and operating expenses. Grants are to be awarded through national competitions to resident management corporations, resident councils, cooperative associations, non-profit organizations and public agencies, which provide $1 for every $4 of federal money. Applicants must replace any housing sold on a one-for-one basis.

Similar types of grants amounting to $331 million over two years are also authorized to promote private ownership of publicly-owned or publicly-held (e.g. distressed properties of the Federal Housing Administration) multi-family buildings. Restoring such properties to good condition offers an opportunity to extend the benefits of home-ownership to low-income families. A third type of home-ownership is also promoted in publicly held, single family properties acquired through defaults and foreclosures. The same two types of planning and implementation grants, amounting to $231 million, are authorized over two years. Families must be first-time buyers and are not required to pay more than 30 per cent of adjusted income to purchase a home.

NAHA's next major plan is an arrangement to preserve the supply of affordable housing that has been threatened by the expiration of restrictions on mortgage pre-payment. Under various subsidy systems governing approximately 360,000 rental housing units built in the late 1960s and early 1970s, private owners were allowed to pay off their loans after 20 years, terminate low-income use restrictions, and convert projects to market rentals, condominiums or other uses. Over the objection of owners, in 1987 Congress imposed a moratorium on pre-payments and conversions until the fall of 1990.

NAHA attempts to reconcile the rights of owners with the continuing need for affordable housing by offering incentives for continued rental to low-income families. The incentive package provides owners with an 8 per cent return on their equity, based on an appraisal of the fair market value. The incentives include additional rent subsidies, increased access to project residential receipts, access to project equity through second mortgages, and higher rents from current tenants. Owners who opt for such a package must maintain their property as affordable rental housing for at least fifty more years. If an owner does not wish to continue renting to low-income households, he must file a notice of intent with HUD and follow a procedure which protects existing tenants, including offering the options of resident home-ownership or of ownership by non-profit, community-based organization. NAHA authorized $425 million in FY 1991 and $838 million in FY 1992.

The fourth NAHA pillar is a new National Homeownership Trust, providing down-payment assistance and interest subsidies to first-time buyers. To qualify, family income can not exceed 95 per cent of the median income of a family of four. The mortgage, with six per cent interest and a minimum down payment of no less than five per cent, may not exceed $124,875.

Another important part of NAHA addresses persons with special needs: the homeless and the elderly. For the homeless, a new Shelter Plus Care programme, with a $382 million authorization over two years, combined housing with supportive services on a one-to-one matching basis. Primarily for those who are seriously mentally ill or substance users, the services include health care, mental health treatment, detoxification, case management education and job training.

The elderly, the fastest growing segment of the Nation's population, are often frail and need supporting services to help them stay in their homes and live more independent and dignified lives, thereby avoiding the more costly institutionalization. A two-year, $90 million Elderly Independence Demonstration Project has been authorized.

In addition, NAHA authorizes three other notable programme innovations: family self-sufficiency; drug elimination; and housing counselling. A comprehensive Family Self-Sufficiency Program is authorized, combining housing assistance (i.e. housing allowances) with services such as job training, child care and transportation to help families become self-sufficient and economically independent.

The programme was voluntary in 1991 and 1992, but became compulsory in 1993 for all Public Housing Authorities.

A two year $327 million authorization is included for a public housing drug elimination programme. Funds may be used to hire security officers, to make physical improvements to enhance security, and to hire people to investigate drug-related crimes. For housing counselling, NAHA authorized $350 million in FY 1991 and $365 million in FY 1992 for a foreclosure-prevention demonstration programme to reduce defaults and foreclosures on FHA-insured, single-family housing.

**An assessment**

Altogether NAHA represents a massive attack on the problems of affordable housing, but appearances may be somewhat deceiving. Unknown in the practice of most industrialized countries, there is in the USA a big difference between authorization and appropriation. Congress and the President may agree on a policy and authorize specific funding for its implementation, but if there is no subsequent appropriation of money, nothing happens. This is the case in the present instance. In February–March 1991, Congress refused to appropriate funds to execute NAHA's 1991 authorizations. Fiscal Year, October 1990–September 1991, will therefore have come and gone with no action taken. As regards an appropriation for NAHA's 1992 authorizations, with the best of intentions, considerable doubts lingered. A non-American may rightly ask: How can this be?

The answer is somewhat complex. In the first place, the newly reached NAHA agreement between Congress and the President and between Democrats and Republicans masks certain continuing deep-seated conflicts on priorities that tend to paralyze follow-up. While the Bush Administration has abandoned Reagan dismantlement, the negative view toward public housing remains. Following the empowerment philosophy, the Bush–Kemp objective is to privatize as much of the public housing stock as possible through the HOPE programme – at the same time paying lip service to the principle that any housing unit sold must be replaced on a one-for-one basis, but with little provision for the funding of such replacement. There is much congressional scepticism and opposition to this approach. The outlook for a HOPE appropriation was clouded further by the release of a study in April 1991, revealing that not only was the show piece of the empowerment policy, the Kennilworth–Parkside project in Washington DC, taking an inordinately long time to rehabilitate, but was also costing around $130,000 per apartment – as compared to an average nationwide HUD cost of renovating a public housing unit of $12,008 and an average HUD cost of rebuilding a new public housing apartment from scratch ranging from $50,000 to $75,000 (Guskind and Steinbach 1991).

Another important political difference concerns subsidy methods. Should the government subsidize the high cost of building new units through producer subsidies (the supply-side approach), or should it subsidize high rents of existing housing units through consumer subsidies (the demand-side approach)? Republicans have strongly favoured housing allowances, i.e. consumer subsidies, on cost grounds and the belief that there is no overall physical shortage of rental housing. Democrats, while believing in the need for consumer subsidies, have strongly maintained that, in addition, persistent local shortages can be met only by subsidizing new housing construction, even though it is more costly per family than housing allowances.

NAHA also contains an interesting bifurcation on home-ownership promotion. While HUD Secretary Kemp seeks to promote home-ownership by selling off public housing, Representative Gonzalez believes that the first-time home-buyer's dream can best be satisfied through the establishment of a National Homeownership Trust which subsidises mortgage down-payments and interest rates. With these and other lesser cross currents in various political agendas, it is not surprising that difficulties have arisen in trying to find proper funding for favourite programmes.

A second explanation of the appropriation impasse is the current 'fixed' limit on available funds. If the only central issue were the comparative need for housing and community development assistance, then a reasonable political accommodation could no doubt be reached on funding programmes covering diverse preferences. However, Congress has been up against three weighty negative considerations: the continuing inability to resolve the federal budgetary deficit problem; the evaporation of the 'Peace Dividend' expected from the ending of the so-called Cold War because of the advent of the Gulf War; and the continuing taxpayers' resistance to higher taxes.

To cope with the deepening budget crisis, in the fall of 1990 Congress reached a Five Year Budget Summit Agreement, which set outside limits for the three major areas of federal spending: domestic discretionary spending; defence; and foreign aid. As a consequence, all domestic discretionary spending, such as health care, environmental protection, space, science, child care, education and housing, would have only $1.5 billion of new money available for FY 1992 appropriations (National Association of Housing and Redevelopment Officials 1991). In view of intense programme competition for such a limited amount of dollars, there was no way that $4 billion could be found in 1992 to fund the new initiatives authorized by NAHA. The FY 1992 outlook for affordable housing was, therefore, not particularly bright.

Nevertheless, in conclusion some positive notes can be sounded. First, notwithstanding its limitations, the new consensus on a comprehensive national housing policy was certainly a landmark in rebuilding national housing programmes (Jacobs 1991, 30). Second, there was the possibility that small amounts of money would in fact be made available in 1992 to at least start some of the newly authorized NAHA programmes. Since a considerable amount of time is required to recruit staff, draft regulations, launch new programmes and get the 'bugs' out, starting slowly on a small scale does have some redeeming features. Third, within another year the economic recession might well be past and tax revenues might be increasing, making possible a more generous funding of new housing programmes in FY 1993. Fourth, positive changes in the political winds of President Bush's 'Kinder, gentler Nation' may well accelerate the process of rebuilding national housing and urban development programmes.

## NOTES

1 1940 data – Department of Housing and Urban Development, *1979 Statistical Yearbook* (Washington DC Government Printing Office 1980), p. 264.
2 Information supplied by Charles Taylor, Director, Division of Budget, Office of Policy Development and Research, Department of Housing and Urban Development, July 20 1990.
3 Information supplied by Herbert G. Persil, Deputy Director, Office of Budget, Department of Housing and Urban Development, 20 July 1990.
4 Among the notable are: Rachel G. Bratt, *Rebuilding a Low-Income Housing Policy* (Philadelphia, Temple University Press 1989); Anthony Downs, *A Strategy for Designing a Fully Comprehensive National Housing Policy for the Federal Government of the United States*, (Cambridge, MA, Massachusetts Institute of Technology 1988); Paul A. Leonard, Cushing N. Dolbeare, Edward Blazere, *A Place to Call Home: The Crisis in Housing for the Poor* (Washington DC, Centre on Budget and Policy Priorities and Low Income Housing Information Service 1989); David G. Schwartz, Richard C. Ferlaute, Daniel N. Hoffman, *A New Housing Policy for America* (Philadelphia, Temple University Press 1988); Raymond J. Struyk, Margery A. Turner, Makiko Ueno, *Future U.S. Housing Policy* (Washington DC, Urban Institute 1988); Irving Welfeld, *Where We Live* (New York, Simon & Schuster 1988).

5 For an ingenious approach to structuring the benefit payment schedule for the housing allowance in the least cost manner, see Irving Welfeld, *Where We Live*, 247–55.
6 George Romney's statement appeared in the *New York Times*, 24 October 1969, as quoted in Henry Aaron, *Shelter and Subsidies*, Washington DC, Brookings Institution 1972, p. 72. For the data on the $2.9 billion deduction, see Aaron, p. 55.
7 The classic discussion of the various options is Joshua E. Greene, *The Tax Treatment of Homeownership: Issues and Options* (Washington DC, Congressional Budget Office, 1981).
8 In defence of tax deductibility, see Susan E. Woodward and John C. Weicher, 'Goring the Wrong Ox: A Defense of the Mortgage Interest Deduction', *National Tax Journal*, September 1989, pp. 301–15.
9 In 1986, the latest available date, out of 56.8 million home-owners (Table 2.7), 29 million took the mortgage interest tax deduction. Internal Revenue Service, *Statistics of Income: Individual Income Tax Returns, 1986* (Washington DC, Government Printing Office 1989), p. 64.

### APPENDIX 1: Housing units subsidized, by tenure and programme 1975–89

**Explanatory notes**
*Column*

1 HUD duplicate reporting, Table 2. Some low-income housing units are assisted by more than one subsidy programme. To take account of this duplication, there must be corresponding adjustment in calculating the total volume of subsidized housing. The duplication factor is derived from columns 1 and 2 of Table 2.
2 Public ownership, Table 2. The number of housing units owned and managed by local Public Housing Authorities operating under the national public housing system.
3 Non-profit ownership (HUD Section 202 programme), data supplied by Assisted Elderly and Handicapped Division, HUD, 8 August 1990. Programme for the elderly and handicapped, enacted in 1959, provides below-market interest rates (generally around 3 per cent) to non-profit sponsors. Since 1974, the programme has been used in conjunction with Section 8 assistance (housing allowances) to reach more low-income families.
4 HUD section 8 programme, Table 2. A housing allowance type of subsidy, enacted in 1974, in which HUD pays the private developer or landlord, directly or through the local Public Housing Authority, the difference between the agreed-upon contract rent based on HUD determined Fair Market Rents, and the

*Appendix 1*: Housing units subsidized, by tenure and programme 1975–89
(millions)

| | | | | Rental housing | | | | | | | Home-ownership | | | |
| | | | | | Private ownership | | | | Tax expenditure* | | HUD | FmHA | Tax expenditure | |
| | | | | | HUD | | | FmHA | | | | | | |
| Year | Total | Less duplicate reporting | Public (HUD) | Non-HUD (Sect. 202) | Sect. 8 | Sect. 236 | Rent Suppl. | Sect. 515 | Tax Exempt Bonds | Tax Credit | Sect. 235 | Sect. 502 | Mortgage Interest Subsidy | Tax Exempt Bonds |
| | | (1) | (2) | (3) | (4) | (5) | (6) | (7) | (8) | (9) | (10) | (11) | (12) | (13) |
| 1975 | – | – | 1.15 | – | – | 0.40 | 0.17 | 0.05 | – | – | 0.41 | – | – | – |
| 1976 | – | – | 1.17 | 0.08 | 0.27 | 0.45 | 0.17 | 0.07 | 0.07 | – | 0.33 | – | – | – |
| 1977 | – | – | 1.17 | 0.10 | 0.46 | 0.54 | 0.18 | 0.10 | 0.12 | – | 0.29 | 0.26 | – | – |
| 1978 | – | – | 1.17 | 0.12 | 0.67 | 0.54 | 0.17 | 0.13 | 0.14 | – | 0.26 | 0.29 | – | – |
| 1979 | – | – | 1.18 | 0.14 | 0.90 | 0.51 | 0.18 | 0.16 | 0.20 | – | 0.23 | 0.31 | – | – |
| 1980 | – | – | 1.19 | 0.16 | 1.15 | 0.54 | 0.16 | 0.19 | 0.06 | – | 0.22 | 0.33 | – | – |
| 1981 | – | 0.16 | 1.20 | 0.17 | 1.32 | 0.54 | 0.16 | 0.21 | 0.03 | – | 0.24 | 0.34 | – | – |
| 1982 | – | 0.18 | 1.22 | 0.19 | 1.53 | 0.54 | 0.15 | 0.24 | 0.12 | – | 0.24 | 0.36 | 24.5 | – |
| 1983 | – | 0.21 | 1.25 | 0.20 | 1.78 | 0.55 | 0.08 | 0.26 | 0.08 | – | 0.23 | 0.40 | 25.6 | – |
| 1984 | – | 0.18 | 1.33 | 0.21 | 1.91 | 0.53 | 0.06 | 0.28 | 0.11 | – | 0.21 | 0.40 | 27.0 | – |
| 1985 | – | 0.20 | 1.35 | 0.22 | 2.01 | 0.53 | 0.05 | 0.30 | 0.58 | – | 0.20 | 0.40 | 28.1 | – |
| 1986 | – | 0.19 | 1.38 | 0.24 | 2.14 | 0.53 | 0.03 | 0.32 | – | – | 0.18 | 0.40 | 29.0 | – |
| 1987 | 34.79 | 0.19 | 1.39 | 0.25 | 2.24 | 0.53 | 0.02 | 0.34 | 0.64 | 0.03 | 0.16 | 0.38 | 28.0 | – |
| 1988 | – | 0.20 | 1.40 | 0.26 | 2.35 | 0.53 | 0.02 | 0.35 | 0.67 | 0.08 | 0.15 | 0.37 | – | (1.0) |
| 1989 | – | 0.20 | 1.40 | 0.27 | 2.42 | 0.53 | 0.02 | 0.37 | 0.68 | – | 0.14 | 0.36 | – | 1.1 |

*Explanatory notes – see following pages*

tenant's rent payment. Until 1981, this was 25 per cent of income, thereafter 30 per cent of income.

5 HUD Section 236 programme, Table 2, Programme adopted in 1968 providing below-market interest rate loans (e.g. 1 per cent) to finance multi-family rental projects by private developers.

6 HUD rent supplement programme, Table 2. Programme enacted in 1965 provided the difference between 25 per cent of the assisted household income and the contract rent targeted for low-income families. Since there were no new commitments after 1973, outstanding obligations decline yearly.

7 Farmers Home Administration Section 515 programme, Table 3. Programme originally adopted in 1949, expanded in 1968, provides reduced-rate loans (as low as 1 per cent) to private developers of rural rental housing.

8 Tax exempt bonds, years 1976–84 – Rasey 1987, 20; years 1985–9. National Council of State Housing Agencies, *Rental Surveys* (1985, 1987, 1988, 1989) (Washington DC), Tables 1, 13. State and local governments are allowed to issue their own, federally tax-exempt bonds, the proceeds of which finance multi-family, private rental housing projects for low- and moderate-income households at below-market interest rates.

9 Tax credits, National Council of State Housing Agencies, *Appendix A, 1987 and 1988, Project Characteristics* (Washington DC, 1989). Low-income housing tax credits, introduced in 1986, are extended to profit and non-profit developers of low- and moderate-income rental projects to cover a percentage (4 or 9 per cent) of qualified expenditures (of the proportion of total costs allocated to the subsidized sector) each year for a 10-year period.

10 HUD Section 235 Programme, Table 2. Programme adopted in 1968, provides below-market interest rates to private lenders on behalf of low- and moderate-income home buyers.

11 Farmers Home Administration Section 502 programme, Table 3. Programme, originally adopted in 1949, provides loans at interest rates as low as 1 per cent to finance home-ownership by low-income farmers.

12 Mortgage interest subsidy, Internal Revenue Service, *Statistics of Income: Individual Tax Returns, 1986* (Washington DC, 1989), p. 64. For years prior to 1986, see IRS Publication No. 1304 by same title in each year. In 1986 (the most recent year available), out of a total of 56.8 million home-owners (Table 8), 29 million (or about 52 per cent) took mortgage interest payments as deductible

items in calculating their federal income taxes. As regards property taxes (on land and house), which were less than one-third
of mortgage interest payments (Table 6), roughly 33 million
home-owners took these deductions.

13 Tax exempt bonds, National Council of State Housing Agencies,
Table 3, *Home Ownership Survey*, 29 January 1991. Cumulative
total, yearly data not available. State and local governments are
allowed to issue federally tax-exempt bonds, the proceeds of
which finance mortgages to moderate-income households at below-
market interest rates because of the federal tax exemption.

* No estimates are available on the number of housing units created
for low- and moderate-income families under the accelerated
depreciation subsidy system (Table 6) for three reasons. First,
there was no reporting of the number of housing units financed
by tax shelter partnerships. Second, the number of tax shelter
partnerships that went into low- and moderate-income rental
housing was comparatively low, around 11 per cent (Verdier
1977, 38). Third, most low- and moderate-income housing financed
by this tax policy would not have been built without the combined
support of other subsidies (Verdier 1977, 45). Therefore, any
attempt to estimate the number of low- and moderate-income
housing units financed would have involved a high degree of
double counting.

## APPENDIX 2: Federal housing subsidies by tenure and type of programme 1975–91

### Explanatory notes
*Column*

1 New public construction. Payment of debt service on bonds issued
to finance public housing up until 1987. Since then costs are
financed by up-front grants.
2 Public housing operating expenses. Special payments to cover
operating deficits of Public Housing Authorities.
3 Non-profit ownership, Section 202 Programme. HUD programme
for the elderly and handicapped, enacted in 1959, provides below-
market interest rate loans (generally around 3 per cent) to non-
profit sponsors.
4 HUD Section 8 programme. A housing allowance type of subsidy
enacted in 1974. HUD pays the private developer or landlord,
directly or through the local Public Housing Authority, the

*Appendix 2:* Federal Housing Subsidies by Tenure and Type of Programme 1975–91
($ billions)

| | | | Rental housing | | | | Private ownership | | | Mixed ownership | | Home-ownership | | | |
| | | | | Non profit | | | HUD | | | | | Tax expenditures | | | |
| | Total | Public New construction | Operating expenses | Section 202 | Section 8 | Rent supp. | Section 236 | Tax expenditures | FmHA (Rural) | HUD Section 235 | Mortgage interest | Property tax | Capital gains | Tax exempt bonds |
| | | (1) | (2) | (3) | (4) | (5) | (6) | (7) | (8) | (9) | (10) | (11) | (12) | (13) |
|---|---|---|---|---|---|---|---|---|---|---|---|---|---|---|
| 1975 | – | 0.97 | 0.34 | – | – | 0.17 | 0.39 | – | 0.18 | 0.19 | – | – | – | – |
| 1976 | 13.60 | 1.27 | 0.63 | – | 0.07 | 0.26 | 0.64 | 0.41 | 0.33 | 0.20 | 4.87 | 4.03 | 0.89 | – |
| 1977 | 13.20 | 1.07 | 0.52 | – | 0.37 | 0.25 | 0.59 | 0.32 | 0.32 | 0.13 | 4.49 | 4.21 | 0.93 | – |
| 1978 | 18.55 | 1.07 | 0.70 | 0.18 | 0.83 | 0.25 | 0.62 | 0.30 | 0.32 | 0.11 | 7.60 | 5.50 | 1.07 | – |
| 1979 | 24.38 | 1.16 | 0.65 | 0.46 | 1.37 | 0.27 | 0.64 | 0.29 | 0.50 | 0.10 | 10.75 | 6.76 | 1.33 | – |
| 1980 | 31.59 | 1.36 | 0.82 | 0.75 | 2.10 | 0.27 | 0.66 | 0.39 | 0.65 | 0.11 | 15.62 | 7.31 | 1.55 | – |
| 1981 | 39.49 | 1.47 | 0.93 | 0.82 | 3.12 | 0.28 | 0.67 | – | 1.11 | 0.20 | 20.15 | 9.13 | 1.61 | – |
| 1982 | 45.53 | 1.77 | 1.01 | 0.74 | 4.09 | 0.28 | 0.67 | 0.40 | 1.51 | 0.26 | 23.31 | 8.36 | 2.22 | 0.91 |
| 1983 | 45.25 | 1.66 | 1.54 | 0.80 | 4.99 | 0.19 | 0.64 | 1.41 | 1.64 | 0.78 | 20.80 | 8.01 | 1.93 | 1.36 |
| 1984 | 49.64 | 1.69 | 1.14 | 0.66 | 6.03 | 0.11 | 0.66 | 1.41 | 2.14 | 0.27 | 22.74 | 8.82 | 2.47 | 1.50 |
| 1985 | 52.07 | 2.20 | 1.21 | 0.50 | 6.82 | 0.07 | 0.62 | 1.56 | 2.72 | 0.16 | 24.79 | 9.32 | 2.61 | 1.69 |
| 1986 | 62.62 | 1.70 | 1.18 | 0.52 | 7.43 | 0.05 | 0.63 | 3.95 | 2.97 | 0.21 | 30.44 | 8.56 | 2.93 | 2.05 |
| 1987 | 68.63 | 0.77 | 1.39 | 0.38 | 8.13 | 0.05 | 0.64 | 1.65 | 3.66 | 0.17 | 34.75 | 10.29 | 4.74 | 2.01 |
| 1988 | 69.40 | 1.04 | 1.49 | 0.30 | 9.13 | 0.05 | 0.63 | 1.71 | 2.68 | 0.18 | 32.68 | 10.10 | 6.64 | 1.77 |
| 1989 | 83.78 | 1.52 | 1.52 | 0.36 | 9.92 | 0.05 | 0.61 | 4.32 | 2.67 | 0.11 | 34.19 | 10.07 | 6.24 | 2.20 |
| 1990 | – | – | – | – | – | – | – | 4.35 | – | – | 39.79 | 11.24 | 16.88 | 2.05 |
| 1991 | – | – | – | – | – | – | – | 4.76 | – | – | 46.60 | 12.43 | 17.55 | 1.88 |

*Explanatory notes – see following pages*

difference between the agreed-upon contract rent (based on HUD determined Fair Market Rents) and the tenant's rent payment. This was 25 per cent of income until 1981, and thereafter 30 per cent.

5 HUD rent supplement programme. Programme enacted in 1965, provided the difference between 25 per cent of the assisted household's income and the contract rent for the targeted low-income families, Since no new commitments were made after 1973, payments are declining.

6 HUD 236 programme. Programme, adopted in 1968, provides below-market interest rate loans (e.g. 1 per cent) to finance multi-family rental projects by private developers.

7 This includes

   i Accelerated depreciation allowance subsidy, years 1976–88 Low Income Housing Information Service (LIHIS), 1989, Table 6: years 1989–91, Office of Management and Budget, (OMB), *Budget of the United States Government, Fiscal Year 1991* (Washington DC 1990), Table C–1, pp. A 71–3. 'Tax shelter' benefits are extended to investment in housing by permitting an accelerated depreciation allowance on the value of the building, so as to create paper losses which can be deducted from ordinary income in calculating federal income taxes. Under the system in effect from 1981 to 1986, buildings could be depreciated over periods of 15, 18 or 19 years, depending on the year the project was actually undertaken. Before 1981, there were no legislatively prescribed depreciable lives, but under Internal Revenue Service guidelines buildings were usually depreciated over 35 to 50 years. These subsidies were terminated in 1986, but outstanding federal commitments continued thereafter on a declining basis (Taylor 1987, 4).

   ii Tax exempt bonds years 1982–8 – LIHIS 1989, Table 6; years 1989–91 – OMB op. cit. Interest income on rental housing bonds issued by state and local governments is exempt from federal taxation. The proceeds are used to finance private multi-family rental housing projects. Tenants must be low- and moderate-income families.

   iii Low-income housing tax credit, years 1987–8 – National Council of State Housing Agencies, *Appendix A, 1987 and 1988 Project Characteristics* (Washington DC, 1989), Table A; years 1989–91 – OMB, op. cit. Tax credits, introduced in 1986, are extended to profit and non-profit developers of low- and moderate-income rental projects to cover percentage (4 or 9 per cent) of qualified expenditures (of the proportion of total costs allocated to the subsidized sector) each year for 10 years.

8 Farmer Home Administration Rural Housing Insurance Fund for losses incurred as a result of below-market interest rates on loans to farmers (Section 502 programme) and to private developers (Section 515 programme).

9 HUD Section 235 programme adopted in 1968, provides below-market interest rates to private lenders on behalf of low- and moderate-income home buyers.

10 Mortgage interest subsidy: years 1976–88 – LIHIS 1989, Table 6; years 1989–91 – OMB, op. cit. Amount of mortgage interest payments deducted from income by home-owners before taxes are calculated.

11 Property tax subsidy, years 1976–88 – LIHIS 1989, Table 6; years 1989–91 – OMB, op. cit. Amount of local property taxes (on land and house) deducted from income by home-owners before taxes are calculated.

12 This includes

   i Capital gains deferred subsidy: years 1976–88 – LIHIS 1989, Table 6; years 1989–91 – OMB, op. cit. Deferral of taxes on capital gains received from home sales, provided that gains are reinvested in another house of equal or greater price within two years after sale.

   ii Capital gains exclusion subsidy: years 1976–88 – LIHIS 1989, Table 6; years 1989–91 – OMB, op. cit. Persons aged 55 or older are granted a one-time exclusion of up to $125,000 capital gain (in income tax calculation) received from sale of their homes. In effect, this provision converts some prior deferrals of tax into forgiveness of tax.

13 Tax exempt bonds: years 1982–8 – LIHIS 1989, Table 6; years 1989–91 – OMB, op. cit. Interest income on mortgage housing bonds issued by state and local governments is exempt from taxation. The bond proceeds are used to finance homes purchased by first-time buyers with low- and moderate-incomes of dwellings with prices under 90 per cent of the average area purchase price.

## REFERENCES

Anderson, M. (1964) *The Federal Bulldozer: A Critical Analysis of Urban Renewal*, Cambridge, MA: MIT Press.

Apgar, W.C., DiPasquale, D., McArdle, N. and Olson, J. (1989) *The State of the Nation's Housing*, Cambridge, MA: Joint Center for Urban Studies, Harvard University.

Arlington County (1990) *Housing Policy Recommendations*, Arlington, Virginia.

Bourdon, R. (1988) *Home Buyer Assistance: Tax-deferred Savings for Downpayments*, Washington DC: Congressional Budget Office.

Bradbury, K.L. and Downs, A. (1981) (eds) *Do Housing Allowances Work?* Washington DC: Brookings Institution.

Bradbury, K.L., Downs, A. and Small, K.A. (1982) *Urban Decline and the Future of American Cities*, Washington DC: Brookings Institution.

Bratt, R.G., Hartman, C. and Myerson, A. (1986) (eds) *Critical Perspectives on Housing*, Philadelphia: Temple University Press.

Bratt, R.G. (1989) *Rebuilding a Low-Income Housing Policy*, Philadelphia: Temple University Press.

Bridge Housing Corporation (1989) *Annual Report*, San Francisco: 32 Second Street, CA 94105.

Brooks, M.E. (1988) *A Survey of Housing Trust Funds*, Washington DC: Center for Community Change, 1000 Wisconsin Ave., NW, 20007.

Burns, L.S. and Grebler, L. (1986) *The Future of Housing Markets*, New York: Plenum Press.

Burt, M. and Cohen, B. (1989) *America's Homeless: Numbers, Characteristics and Programs That Serve Them*, Washington DC: Urban Institute.

Caprara, D. and Alexander, B. (1989) *Empowering Residents of Public Housing*, Washington DC: Center for Neighborhood Enterprises, 367 Conn Ave., NW.

Clay, P.L. (1987) *At Risk of Loss: The Endangered Future of Low-Income Rental Housing Resources.* Washington DC: Neighbourhood Reinvestment Corporation.

Council of Economic Advisors (1990) *Economic Report of the President*, Washington DC: Government Printing Office.

Department of Housing and Urban Development (1980) *1979 Statistical Yearbook*, Washington DC: Government Printing Office.

—— (1981) *Rental Housing: Condition and Outlook*, Washington DC: Government Printing Office.

—— (1984) *Report to the Secretary on the Homeless and Emergency Shelters*, Washington DC: Government Printing Office.

—— (1987a) *Official U.S. Special Merit Award Projects*, Washington DC: Government Printing Office.

—— (1987b) *A Directory of Official U.S. IYSH Projects*, Washington DC: Government Printing Office.

—— (1987c) *Affordable Residential Construction: Challenge and Response: A Guide for Home Builders*, Washington DC: Government Printing Office.

—— (1988) *The President's National Urban Policy Report*, Washington DC: Government Printing Office.

—— (1989a) *Annual Housing Survey for the United States in 1987*, Washington DC: Government Printing Office.

—— (1989b) *The 1988 National Survey of Shelters for the Homeless*, Washington DC: Government Printing Office.

—— (1989c) *Homeless Assistance Policy and Practice in the Nation's Five Largest Cities*, Washington DC: Government Printing Office.

—— (1990a) *HOPE: Homeownership and Opportunity for People Everywhere*, Washington DC: Government Printing Office.

—— (1990b) *Report to Congress on Alternative Methods of Funding Public Housing Modernization*, Washington DC: Government Printing Office.

—— (1991a) *Homeownership and Affordable Housing: The Opportunities*, Washington DC: Government Printing Office.

—— (1991b) *President's Report on National Urban Policy: Recapturing the American Dream*, Washington DC: Government Printing Office.

Dolbeare, C.N. (1990) *At a Snail's Pace*, Washington DC: Low Income Housing Information Service.

Downs, A. (1983a) *Rental Housing in the 1980s*, Washington DC: Brookings Institution.

—— (1983) *The Revolution in Real Estate Finance*, Washington DC: Brookings Institution.

Downs, A. and Small, K.A. (1982) *Urban Decline and the Future of American Cities*, Washington DC: Brookings Institution.

Dreier, P., Schwartz, D.C. and Greiner, A. (1988) 'What Every Business Can Do About Housing', *Harvard Business Review*, September–October 1988.

Enterprise Foundation (1989) *Doing What it Takes*, Columbia, Maryland, 505 American City Building, 21044.

Fricker, M. and Pizzo, S. (1990) 'Bailing Out a Sinking Industry', *Public Citizen*, July–August 1990: 6–8.

Frieden, B.J. and Kaplan, M. (1975) *The Politics of Neglect*, Cambridge, MA: MIT Press.

Friedman, J. and Weinberg, D.H. (1983) (Eds) *The Great Housing Experiment*, Beverley Hills, CA: Sage Publications.

Gale, D.E. (1984) *Neighborhood Revitalization and the Postindustrial City*, Lexington, MA: Heath.

General Accounting Office (1979) *Rental Housing: A National Problem That Needs Immediate Attention*, Washington, DC: Government Printing Office.

—— (1989) *Housing Conference: National Housing Policy Issues*, Washington, DC: Government Printing Office.

Goetze, R. (1983) *Rescuing the American Dream: Public Policies and the Crisis in Housing*, New York: Holmes and Meiar.

Gravelle, J.G. (1987) *Tax Policy and Rental Housing: An Economic Analysis*, Washington DC: Congressional Research Service.

Greene, J. (1981) *The Tax Treatment of Homeownership: Issues and Options*, Washington DC: Congressional Budget Office.

Guskind, R. and Steinbach, C. (1991) 'Sales Resistance', *National Journal*, 6 April 1991: 798–803.

Habitat for Humanity (1990) *Habitat World*, Americus, GA. 31709.

Hartman, C. (1986) 'Housing Policies Under the Reagan Administration' in Bratt, R.G., Hartman, C. and Meyerson, A. (1986) (Eds) *Critical Perspectives in Housing*, Philadelphia, PA: Temple University Press, pp. 362–76.

Housing Assistance Council (1989) *Annual Report*, Washington DC: 1025 Vermont Ave., NW, 20005.

Howenstine, E.J. (1983) *Attacking Housing Costs: Foreign Policies and Strategies*, New Brunswick, NJ: Center for Urban Policy Research, Rutgers University Press.

—— (1985) *Converting Public Housing to Individual and Cooperative*

*Ownership: Lessons from Foreign Experience.* Washington DC: Department of Housing and Urban Development, HUD USER RP 3213.

—— (1986) *Housing Vouchers: A Comparative International Analysis*, New Brunswick, NJ: Center for Urban Policy Research, Rutgers University Press.

Hughes, J.W. and Sternlieb, G. (1987) *The Dynamics of America's Housing*, New Brunswick, NJ: Center for Urban Policy Research, Rutgers University Press.

Internal Revenue Service (1989) *Statistics of Income Individual Income Tax Returns*, Washington DC: Government Printing Service.

Jacobs, B.G. (1991) 'Landmark Legislation Offers Housing Solutions', *National Real Estate Investor*, January 1991, p. 30.

Kennedy, S.D. and Finkel, M. (1987) *Report of First Year Findings for the Freestanding Housing Voucher Demonstration*, Washington DC: Department of Housing and Urban Development.

Leonard, P.A., Dolbeare, C. and Blazere, E. (1989) *A Place to Call Home: The Crisis in Housing for the Poor*, Washington DC: Center on Budget and Policy Priorities and Low Income Housing Information Services.

Liner, E.B. (1989) *Decade of Devolution: Perspectives on State and Local Relations*, Washington DC: Urban Institute.

Local Initiatives Support Corporation (1989) *Annual Report*, New York: 666 Third Ave., 10017.

Low Income Housing Information Service (1989) *Special Memorandum: The Fiscal Year 1990 Budget and Low Income Housing*, Washington DC: 1012 14th St., NW. 20005.

—— (1990) *Expiring Section 8 Contracts: Renewing the Promise of Affordable Housing*, Washington DC: 1012 14th St., NW, 20005.

McGeary, M.G.H. and Lynn, L.E. (1988) (Eds) *Urban Change and Poverty*, Washington DC: National Academy Press.

Mannheim, U. (1981) *Rental Housing in the 1980s*, Washington DC: National Association of Home Builders.

Mayo, S.K., Mansfield, S., Warner, M.D. and Zwetchkenbaum, R. (1980) *Housing Allowances and Other Rental Assistance Programs – A Comparison Based on the Housing Allowance Demand Experiment. Part 2 – Costs and Efficiency*, Cambridge, MA: Abt Associates.

Miles, B.L. (1988) *The Housing and Community Development Act and the Stewart S. McKinney Homeless Assistance Act: Summary and Analysis*, Washington DC: Congressional Research Service.

—— (1990) *Housing Policy: Homeownership Affordability*, Washington DC: Congressional Research Service.

Milgram, G. (1990) *Trends in Funding and Numbers of Households in HUD-Assisted Housing. Fiscal Years 1975–1990*, Washington DC: Congressional Research Service.

Muth, R.F. (1981) 'Commerce and Housing Credit' in McAllister, E.J. (Ed.) *Agenda for Progress: Examining Federal Spending*, Washington DC: Heritage Foundation, pp. 143–58.

National Association of Housing and Redevelopment Officials (1991) *Legislative Update*, 26 April 1991.

National Housing Task Force (1988) *A Decent Place to Live*, Chairman: James M. Rouse; Vice Chairman: David O, Maxwell (Rouse/Maxwell Report) Washington DC: 1625 Eye St., NW, Suite 1015, 20006.

## 66   The new housing shortage

National Low Income Housing Coalition (1986) *Low Income Housing Policy Statement*, Washington DC: 1012 14th St., NW, 20005.

—— (1990) *Roundup*, No. 130, March 1990, Washington DC.

National Low Income Housing Preservation Commission (1988) *Preventing the Disappearance of Low Income Housing*, Washington DC: Balmat; Co-Chairs: Charles A. Mills and Henry S. Reuss.

Neighbourhood Reinvestment Corporation (1989) *Annual Report*, Washington DC: 1325 G. St., NW, Suite 800, 20005.

Pedone, C. (1988) *Current Housing Problems and Possible Federal Responses*, Washington DC: Congressional Budget Office.

Peirce, N.R. and Steinback, C.E. (1987) *Corrective Capitalism: The Rise of Community Development Corporations*, New York: Ford Foundation.

Peterson, G.E. and Lewis C. (1986a) (Eds) *Reagan and the Cities*, Washington DC: Urban Institute.

—— (1986b) (Ed.) *The Reagan Block Grants: What Have We Learned?* Washington DC: Urban Institute.

President's Commission on Housing (1982) *Report*, Washington DC: Government Printing Office; Chairman: William F. McKenna; Vice Chair: Carla A. Hills.

Rasey, K.P. (1987) *The Financing of Rental Housing*, Washington DC: Congressional Research Service.

Rohe, W. and Stegman, M.A. (1990) *Public Housing Homeownership Demonstration Assessment*, Washington DC: Department of Housing and Urban Development.

Rosen, K.T. (1984) *Affordable Housing: New Policies for the Housing and Mortgage Markets*, Cambridge, MA: Ballinger.

Saunders, L. (1989) 'House-Hunting? Read This First', *Forbes*, 20 March 1989, pp. 119–21.

Scanlon, J. (1990) *People Power in the Projects: How Tenant Management Can Save Public Housing*, Washington DC: Heritage Foundation.

Schussheim, M.J. (1988) *Summaries of Papers on U.S. Housing Policy prepared for the Center for Real Estate Development, The Massachusetts Institute of Technology*, Washington DC: Congressional Research Service.

—— (1989) *The Proposed National Affordable Housing Act: Summary and Evaluation*, Washington DC: Congressional Research Service.

—— (1990) *U.S. Housing: Problems and Policies*, Washington DC: Congressional Research Service.

Stegman, M.A. and Sumka, H. (1976) *Nonmetropolitan Urban Housing: An Economic Analysis of Problems and Policies*, Cambridge, MA: Ballinger.

Stegman, M.A., Sumka, H. and Holden, D.J. (1987) *Nonfederal Housing Programs*, Washington DC: Urban Land Institute.

Struyk, R.J. and Bendlick, M., Jr. (1981) (Eds) *Housing Vouchers for the Poor: Lessons From a National Experiment*, Washington DC: Urban Institute.

Struyk, R.J., Bendlick, M., Jr., Mayer, N. and Tuccillo, J.A. (1983) *Federal Housing Policy at President Reagan's Midterm*, Washington DC: Urban Institute.

Struyk, R.J., Bendlick, M., Jr., Turner, M.A. and Ueno, M. (1988) *Future U.S. Housing Policy*, Washington DC: Urban Institute.

Sundquist, J.L. and Davis, D.W. (1969) *Making Federalism Work*, Washington DC, Brookings Institution.

Taylor, J. (1987) *Income Tax Treatment of Rental Housing and Real Estate Investment After the Tax Reform Act of 1986*, Washington DC: Congressional Research Service.

Turner, M.A. and Reed, V. (1990) *Housing America: Learning From the Past, Planning for the Future*, Washington DC: Urban Institute.

Vanhorenbeck, S.M. (1989) *Housing for the Elderly and Handicapped: Section 202*, Washington DC: Congressional Research Service.

Verdier, J.M. (1977) *Real Estate Tax Shelters and Direct Subsidy Alternatives*, Washington DC: Congressional Budget Office.

Weicher, J.C. (1980) *Federal Policies and Programs*, Washington DC: American Enterprise Institute.

Weicher, J.C., Villani, K.E. and Roistacher, E. (1981) *Rental Housing: Is There a Crisis?* Washington DC: Urban Institute.

Welfield, I. (1988) *Where We Live*, New York: Simon and Shuster.

Wilson, J.Q. (1966) (Ed.) *Urban Renewal: The Record and the Controversy*, Cambridge, MA: MIT Press.

Woodward, S.E. and Weicher, J.C. (1989) 'Goring the Wrong Ox: A Defense of the Mortgage Interest Deductibility', *National Tax Journal*, September 1989, pp. 301-15.

Zigas, B. (1989) *Low Income Housing and Homelessness: Facts and Myths*, Washington DC: Low Income Housing Information Service.

# 3 Housing tenure and affordability

## The British disease

*Peter Malpass*

The main theme of this chapter is the dominance of tenure considerations in British housing policy, most obviously since 1979. By making housing tenure seem a mark of social success or failure, 'tenure policy' has contributed to the emergence of an acute shortage of affordable housing. The Government's emphasis on the virtues of home-ownership, and the consequent reduction in the supply of cheap rented housing, has meant that low-income households now find it difficult to obtain suitable accommodation. Unable to afford the inflated selling prices asked for even the most modest houses, they find that rented accommodation is a scarce and distant alternative. The problem of affordable housing is particularly marked in the South. There are now large regional variations in house prices, rents and availability, which have important implications for labour mobility. Underlying the very real problems of affordability is a deep confusion in housing policy about what constitutes affordable housing.

The contradictions of the British housing system lie in the inequitable treatment of different tenures, reflecting official preference for owner-occupation. The general level of rents is considered to be too low, despite the low, or very low, incomes of most tenants, and there have been several legislative attempts to raise rent levels, while providing more 'targeted assistance' with housing costs. Meanwhile, many recent, heavily-mortgaged purchasers have found themselves squeezed by a combination of very high prices in relation to earnings, and record interest rates. The tax privileges afforded to home-owners are both hugely expensive and poorly targeted, yet there has been no attempt at reform, and none is likely in the foreseeable future.

## HOUSING TENURE AND THE HOUSING PROBLEM

This chapter has three sections. The first looks at the nature of housing tenure, and recent changes in housing tenure in Britain. This is followed by a review of housing policy since 1979 and the current affordability crisis. The final section brings together the main points and looks to future prospects for change.

Before attempting to compile empirical evidence of the 'housing problem', we must ask what such a problem would consist of and how it could be measured. The need for shelter is inherent in the human condition. The most basic questions are therefore: is there enough housing to meet the needs of the population, and is it of a satisfactory standard? These questions cannot be answered in absolute terms; measures of quantity and quality are contingent upon wider social conditions and expectations. Moreover, in societies where housing provision is based on individual self-help (as in some Third World countries) homelessness is unlikely to be a serious problem but quality will be restricted by the skill and other resources available to self-builders.

In more developed market economies, housing becomes a commodity to be produced for its exchange value rather than its use value. This changes the nature of the problem. First, it means that consumers' access to suitable housing is determined by their ability to pay, rather than their construction skills; price and affordability replace physical strength and skill as key determinants of housing consumption. Second, it means that producers and consumers are now different groups, with different interests and perspectives. Thus, following Ball (1983, 1986), it is important to recognize that the housing problem in capitalist countries includes both the difficulties faced by consumers in securing adequate accommodation in a suitable location at a price they can afford, and the problems faced by producers in securing a market for their products.

Housing is inherently expensive to produce, and is generally too expensive for most people to buy outright, especially in their early adult lives; thus some form of loan finance is needed. It is important to note that producers and consumers have different interests. For house-builders, profitability and continued business viability depend on the rate at which capital invested in new production can be realized, or invested in the development of the next site. House-builders thus need to sell as quickly as possible, but consumers need some method of spreading expenditure over time. These differing interests give rise to housing problems of a different type. The production of housing as a commodity can solve the basic shelter problem, but it gives rise to a new problem: affordability.

Developed economies have developed two main ways of financing housing consumption: various forms of renting, and mortgaged house purchase. Historically, the dominant system was private renting, in which people with capital acquired a long-term interest in the newly built properties. This allowed the builder's capital to be realized, and enabled tenants to occupy dwellings for a regular rent, part of which represented interest on the capital tied up in the property. In Britain in the twentieth century, local authorities and, to a lesser extent, housing associations have largely replaced private landlords, but the essence of the system remains the same: a long-term recycling of capital is undertaken by an agency standing between the producer and the consumer. With mortgaged individual home-ownership, there is no intermediary agency; consumers themselves raise and repay long-term loans. The crucial task of building societies and other financial institutions is to provide these long-term loans to individuals.

How do the financing systems work in practice, and how can their effectiveness be evaluated? They can be said to work satisfactorily if housing suppliers are able to sustain their operations, and meet their corporate objectives, and if consumers can acquire suitable accommodation at an affordable price. It can be argued that a certain level of bankruptcies on the one hand, and homelessness or affordability difficulties on the other, should be regarded as normal. The most serious arguments arise when homelessness and affordability problems reach levels which call into question the effectiveness of the prevailing systems.

### Housing tenure in Britain

The development of housing studies in Britain has taken place within a framework based on housing tenure categories. However, tenure labels can obscure important features of the landscape, and must be handled with care. The emphasis on tenure in this chapter reflects the overwhelming importance attached to home-ownership in government policy rather than a commitment to tenure analysis as such.

Tenures are essentially consumption categories, describing the way households pay for housing, and their relationship to its ownership (Ball 1986). But, especially in Britain, tenures have become associated with particular forms of production and distribution, and with different social images: the dominant image of owner-occupation is very different from that of local authority renting. This reflects and reinforces social differentiation along tenure lines. The four main tenure categories in Britain are: owner-occupation, local authority renting, private renting, and housing association renting. There are additional small categories covering cooperatives, and 'non-tenure housing' (Whitehead and Kleinman 1988), i.e. squatting and informal sharing.

Tenures are convenient labels, although they should not be allowed to obscure the great diversity within categories (Forrest, Murie and Williams 1990; Allen and MacDowell 1988). Owner-occupation currently accounts for two-thirds of all housing in Britain; local authority renting is the next largest tenure, at about a quarter. Private and housing association renting together make up less than ten per cent of the total. This pattern is the outcome of a major restructuring of housing tenure during the twentieth century. Before the First World War about 90 per cent of housing was privately rented, and only about 10 per cent was owner-occupied. Provision by local authorities and housing associations was negligible. Since then, there has been a steady decline in private tenancy, accompanied by a steady rise in owner-occupation and, until 1979, in local authority housing. Since 1979, the size of the local authority sector has fallen steadily (Figure 3.1). This fall in the amount of 'social housing' has not been compensated by a rise in the amount of housing rented from housing associations, which still accounts for only 3 per cent of the stock.

A feature of the change over the past thirty years or so has been a marked increase in social polarization by tenure (Forrest and Murie 1988, Willmott and Murie 1988). Surveys consistently show a clear and increasing preference for owner-occupation over local authority and private renting. In 1967, 66 per cent of people surveyed said that their preferred tenure was owner-occupation, compared with 23 per cent who preferred local authority renting and just 11 per cent who preferred private renting (Building Societies' Association 1983 p. 10). By 1986, the corresponding figures were 77 per cent, 11 per cent and 4 per cent, while no fewer than 93 per cent of people aged 25–34 *expected* to be home-owners within 10 years (BMRB 1986, Table 11). The cultivation of these preferences has played an important part in tenure restructuring, and has been very successful; local authority housing has become a second choice, or even a last resort, for many people.

While owner-occupation has become increasingly socially diverse as it has grown, local authority housing has experienced a growing concentration of the least well-off. Family Expenditure Survey data show that in 1953–4 only 16 per cent of families in the bottom quarter of the income distribution were local authority tenants, compared with 43 per cent in 1976 (Gray 1979, p. 201). The percentage of local

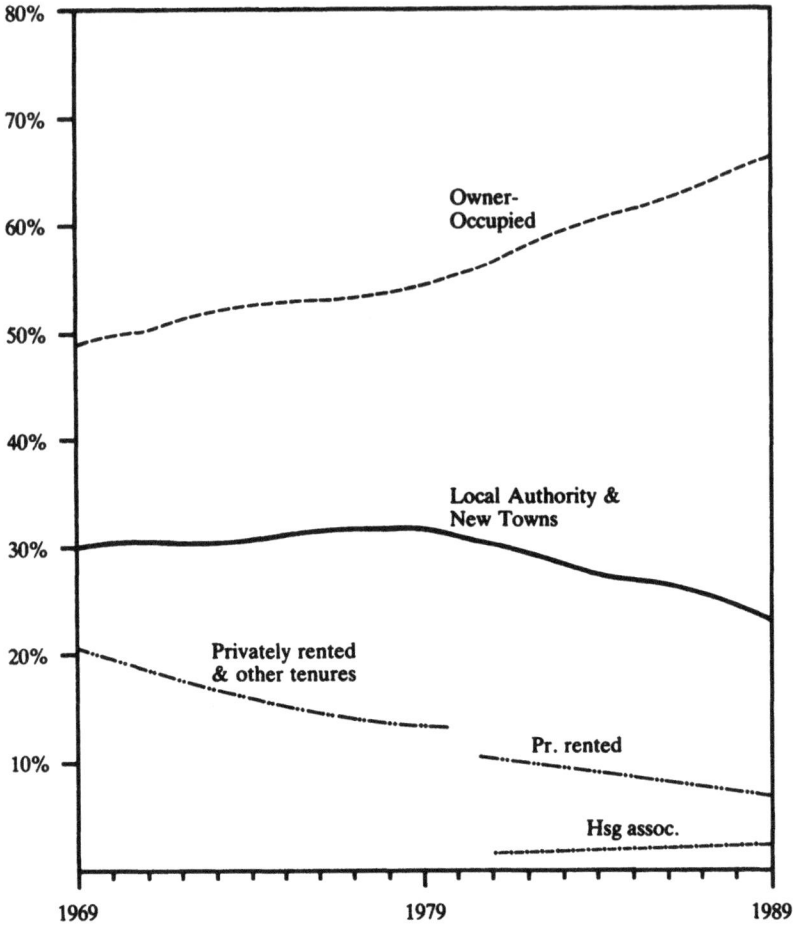

*Figure 3.1*    Housing tenure in Great Britain 1969–89
*Source:    Housing and Construction Statistics*, HMSO

authority tenants who were in the bottom decile rose from 32.5 per
cent in 1968 to 56.1 per cent in 1978 (Robinson and O'Sullivan 1983).
Bentham (1986) also found evidence of the widening income gap
between local authority tenants and home-owners. In 1953, the
median income of council tenants was 89 per cent of that for home-
owners, but had fallen to 45 per cent in 1983, despite a substantial
increase in the numbers of working class owner-occupiers.

    More recent FES data confirms these trends; in 1986 58.9 per cent
of tenants were in the lowest three income deciles, compared with

just 14 per cent of home-owners. In 1987, local authority tenants had a median income of 47.1 per cent of the median income of all households and only 30.6 per cent of that for mortgaged home-owners. In 1988, two-thirds of council tenants had weekly incomes of less than £150, compared with 18 per cent of owners.

Hamnett's (1984) study of tenure differences in terms of socio-economic groups provides an alternative approach. He concluded that,'Skilled manual workers have moved increasingly into the owner-occupied sector, leaving behind them a council sector increasingly dominated by the semi-skilled, the unskilled and the economically inactive'. In the period 1961–81 there was a three-fold increase in the proportion of households counted as economically inactive in the population as a whole, but in the local authority sector the increase was more than five-fold, from 4.8 per cent to 28 per cent. By 1987, 56 per cent of all local authority renting households contained no wage or salary earners, compared with 32 per cent for all tenures together.

Two significant factors here are the high incidence of unemployment amongst local authority tenants and the increase in the proportion of elderly tenants. In 1981 local authority housing in England and Wales contained just over 28 per cent of all dwellings, but 36 per cent of all 'household heads' over the age of 60. It has recently been claimed that retired people provide up to a quarter of new local authority tenants (Pickup 1986). In addition to the elderly, people under 25 have become over-represented in local authority housing, while people aged 39 to 59 are under-represented (*Social Trends 1988*).

Other disadvantaged groups have become concentrated in local authority housing. Forrest and Murie (1988 p. 68) have brought together social security figures showing that, between 1967 and 1984, the proportion of Supplementary Benefit claimants who were council tenants rose from 45 per cent to 61 per cent, whereas the proportion who were owner-occupiers only rose from 17 per cent to 21 per cent, despite the considerable growth in home-ownership. Forrest and Murie also looked at the proportion of Supplementary Benefit claimants in the two tenures. Between 1979 and 1984 the proportion of home-owners claiming benefit rose from 4 per cent to 5 per cent, while the proportion of local authority tenants rose from 21 per cent to 34 per cent. A final source of information about the composition of local authority tenants is the *London Housing Survey 1986–87*, which showed that 4 per cent of all London households were headed by single parents, but that in the local authority sector the figure was

9 per cent (London Research Centre 1988). Nearly half of Afro-Caribbean households in London were in council housing, and this group was under-represented in owner-occupation.

All these indicators reflect the tendency for local authority housing to become the tenure of the least well-off and others who are socially and economically disadvantaged. This is not to imply that all local authority tenants are poor or that there are no poor home-owners. There is, indeed, very considerable variation in the incomes of owners, both within and between regions. The dominant feature of owner-occupation today is its heterogeneity, embracing both rich and poor. Local authority housing, on the other hand, is becoming increasingly homogeneous, with fewer and fewer tenants on above average earnings and an increasing proportion which is outside the workforce altogether.

### Housing policy and the obsession with tenure

For twenty-five years after the Second World War, British housing policy was dominated by issues of quantity and quality. There were more households than dwellings, and there was a serious problem of unfit housing inherited from the past. The emphasis in policy varied from time to time, but all governments pursued high output policies, and accepted that local authorities had a major contribution to make by building houses, both to reduce the shortage and to replace unfit housing. By the late 1960s, however, Ministers were beginning to signal an end to high output policies, and new preoccupations emerged. The 1970s became a decade when financial issues and debates came to prominence, and with the election of the first Thatcher administration in May 1979 a new phase began. The 1980s were dominated by tenure obsessions which, it is argued here, have contributed to the current housing crisis.

The political context within which housing policy has developed since 1979 has been one in which an aggressive, radical, right-wing government has pressed forward with a political philosophy emphasizing the limitations of government, the disadvantages of tax-based public expenditure and the strengths and virtues of private enterprise. A feature of the Thatcherite approach has been a particularly severe and sustained attack on local government in general, and local authority housing in particular.

More generally, the Thatcher Governments were prepared to move rapidly towards the implementation of radical policies, showing scant regard for reasoned criticism of their actions and a dismissive attitude to the idea of consensus politics. The stamp of Thatcherite government was an emphasis on economic liberalism and deregulation on the one hand, and a huge increase in the power of central government on the other hand.

The economic context has been one of deep recession in the early 1980s, with very high levels of unemployment and business failure, followed by faltering recovery, and renewed recession. The period since 1979 has been one of very high interest rates, which have had important implications for the housing market. Whereas in the 1970s governments sought to keep interest rates down, and showed a readiness to intervene specifically to protect home-buyers from the impact of wider interest rate rises, the new Conservative Government allowed the mortgage interest rate to reach 15 per cent in 1979, and again in 1981, without feeling constrained to step in to help mortgage payers. The combined effect of high interest rates and economic recession was that private house building reached a twenty-five year low point in 1981; activity later recovered slightly, before falling to a new low in 1989 (Figure 3.2).

The housing market was volatile and subject to structural change during the 1980s. In real terms, house prices fell in 1981 and 1982 (Hills forthcoming, Figure 4.2) but by 1988 there was a near record level of overall increase, recorded at 32 per cent (Nationwide Anglia Building Society 1989). In addition to this chronic instability in prices, the housing market was shaken up by changes affecting the construction industry (Ball 1988) and a major restructuring of mort gage lending and estate agency services. The building societies faced a strong challenge from the banks, and sought new powers to operate more competitively (Boddy 1989). One of the leading societies, the Abbey National, actually turned itself into a bank. The new climate was also reflected in the widespread acquisition of estate agency businesses by building societies, banks and insurance companies.

## THE THREE PHASES OF GOVERNMENT POLICY

The development of housing policy since 1979 has taken place in three phases, broadly coinciding with the three Thatcher Governments. In the first phase the new Conservative administration set about promoting the growth of home-ownership, to the virtual exclusion of other objectives. During the mid-1980s there was a period of consolidation and implementation, or drift and uncertainty, depending on one's point of view. The third phase began in the run up to the general election of 1987 and went on to include major

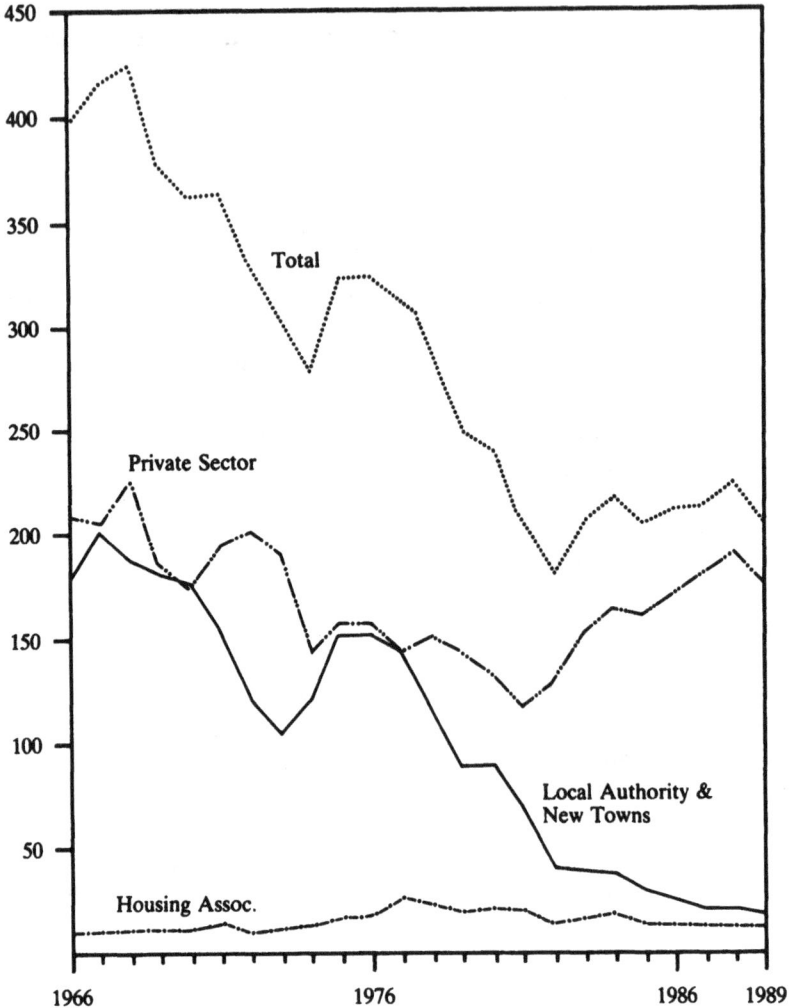

*Figure 3.2*   Permanent dwellings completed in the UK 1966–89 ('000's)
*Source:   Housing and Construction Statistics*, HMSO

legislation in the Housing Act 1988 and the Local Government and
Housing Act 1989.

The broad policy objectives for the 1980s were set out by Michael
Heseltine, the newly appointed Secretary of State for the Environment. Addressing the Institute of Housing Conference in June 1979,
he identified four objectives:

1 to increase individual freedom of choice and sense of personal
   opportunity. This, he explained, meant expansion of opportunities
   for home-ownership, claiming: 'Dreams are going to come true for
   many more people';
2 continuing improvement in the quality of housing;
3 greater value for money (to be achieved partly by removing
   restraints on private house builders);
4 better use of resources, by concentrating them where housing
   needs were most acute.

Notably absent from this and subsequent statements of policy
objectives have been references to building targets and homelessness.
Heseltine pointedly refused to estimate future housing requirements,
preferring to concentrate on 'what the country can afford', and the
need to create a climate in which private builders could flourish.

The most radical measure to promote owner-occupation contained
in the Housing Act 1980 was the right for secure local authority
tenants to buy their homes at substantial discounts from the market
price, and with automatic right to a local authority mortgage, if
required. Table 3.1 shows the rate of sales from 1979–88, including
both sales under the right to buy and other disposals to individual
owners.

*Table 3.1* Sales of public sector dwellings in Great Britain 1979–88

|      | Local authorities | New towns | Housing associations | Total   |
|------|-------------------|-----------|----------------------|---------|
| 1979 | 42,285            | 1,685     | –                    | 43,970  |
| 1980 | 85,840            | 6,350     | –                    | 92,190  |
| 1981 | 112,061           | 5,645     | 7,599                | 125,305 |
| 1982 | 215,797           | 6,675     | 18,221               | 240,693 |
| 1983 | 157,382           | 7,225     | 17,343               | 181,950 |
| 1984 | 118,454           | 6,270     | 12,591               | 137,315 |
| 1985 | 106,268           | 4,685     | 9,359                | 120,412 |
| 1986 | 101,068           | 4,065     | 8,079                | 113,212 |
| 1987 | 122,319           | 5,329     | 7,204                | 134,852 |
| 1988 | 176,518           | 7,321     | 8,529                | 192,368 |

*Source:* Housing and Construction Statistics, HMSO

Discounts under the 1980 Act began at 33 per cent for a tenant of
three years' standing, and rose to 50 per cent for tenants of twenty
years. It is hardly surprising that sales were initially very high, given
that the Government used other powers to increase local authority
rents very rapidly in 1981 and 1982 (Malpass 1990).

However, the rate of sales then declined for five consecutive years,

*Table 3.2* Dwellings completed in Great Britain 1979–89

|      | Local authorities | New Towns & government departments | Housing associations | Private builders | Total |
|------|------|------|------|------|------|
| 1979 | 75,573 | 10,615 | 17,835 | 140,481 | 244,504 |
| 1980 | 76,997 | 9,030 | 21,097 | 128,139 | 235,263 |
| 1981 | 54,889 | 10,619 | 19,306 | 114,769 | 199,583 |
| 1982 | 33,205 | 4,033 | 13,070 | 124,933 | 175,241 |
| 1983 | 32,805 | 2,292 | 16,092 | 147,218 | 198,407 |
| 1984 | 31,593 | 2,355 | 16,642 | 158,721 | 209,311 |
| 1985 | 26,125 | 1,102 | 13,069 | 155,027 | 195,323 |
| 1986 | 21,277 | 1,292 | 12,294 | 166,858 | 201,721 |
| 1987 | 18,671 | 1,280 | 11,607 | 174,175 | 205,734 |
| 1988 | 18,828 | 764 | 11,215 | 184,767 | 215,574 |
| 1989 | 15,107 | 886 | 11,322 | 169,199 | 196,514 |

*Source:* Housing and Construction Statistics, HMSO

before being revived by still higher levels of discounts, reaching 70 per cent for tenants for fifteen years' standing who buy flats (the maximum discount on houses is 60 per cent, but only after thirty years' eligible tenancy). The much higher level of sales in 1988 probably reflects tenants' concerns about possible transfers to alternative landlords, as provided for in the Housing Act 1988. The sales figures should be examined alongside the construction statistics for the 1980s (Table 3.2).

These tables show that sales exceeded new building by the private sector in both 1981 and 1982, indicating that the sale of public sector dwellings was the major source of growth in the owner-occupied sector during the worst of the economic recession in the period 1981–7.

The rate of building by local authorities fell steadily during the 1980s, reaching the lowest levels since the introduction of subsidies in 1919. Still lower levels of new building are planned for the early 1990s, down to as few as 6,000 units (Treasury 1989, Table 9.8). Sales by councils far exceed their new building in every year from 1980 onwards. Until 1979, the total stock of public housing increased every year; since 1979, it has decreased every year. Local authority and new town housing in Great Britain peaked at just over 6.5 million dwellings in 1979, and by the end of 1989 had fallen to 5.3 million, a reduction of 18.3 per cent.

Alongside council tenants' 'right to buy', the Government launched a series of other initiatives to promote home-ownership among low-income households but, compared to the 'right to buy' the impact was very limited (Forrest, Lansley and Murie 1984; Booth and Crook 1985). The success of the campaign to sell council houses was due to attractive discounts and rapidly rising local authority rents. The Housing Act 1980 introduced a new subsidy system which gave the Secretary of State powerful leverage on rent levels: by calculating actual subsidy in accordance with notional rent increases it was possible to put authorities into a position where there was very little choice but to raise rents (Malpass 1990). An important feature of the 1980 subsidy system was that rent increases were engineered by withdrawal of subsidy, with the result that tenants were put in a position where higher rents were not directly reflected in higher standards of service. In particular, it can be argued that the Government was the 'agent of disrepair' to the extent that it reduced local authority housing subsidy by more than 80 per cent in three years in the early 1980s. The weakness of the new subsidy system was that, although it permitted the centre to force up rents, it could only do so by reducing subsidy; there was no mechanism for channelling extra resources into repair and maintenance expenditure.

Nevertheless, the Conservative approach to social policy emphasizes the channelling of assistance to those in greatest need, and there was legislation to improve targeting of financial assistance towards tenants on the lowest incomes. The Social Security and Housing Benefits Act 1982 introduced reforms in the payment of means-tested rent assistance in a way designed to redistribute resources away from tenants on moderate incomes towards those on the lowest incomes. No doubt some tenants interpreted this as yet another incentive to exercise their right to buy: with rents rising and entitlement to assistance declining, the prospect of home-ownership, at a discount price, and with unquestioned entitlement to mortgage interest relief, must have seemed more attractive to some.

The introduction of Housing Benefit was, strictly speaking, a social security measure, and there was no major Housing Act between 1980 and 1988. The Housing Defects Act 1984, the Housing and Building Control Act 1984, and the Housing and Planning Act 1986, were about fine tuning and damage limitation rather than major reform (Malpass and Murie 1990, Chapter 5).

In 1987, the third Thatcher administration gathered itself for a further bout of reform. Whereas the emphasis in the early 1980s was heavily weighted towards the promotion of home-ownership, by the third term there was a discernible change of emphasis; without diminishing its support and enthusiasm for owner-occupation, the Government turned its attention to the rented sectors (Kemp 1989).

The new approach took the form of a vigorous assault on local authority housing, which was characterized by Ministers as a 'failed solution'. The importance of this attack should not be underestimated, for it marked a real departure from the attitudes of all previous governments since the Second World War. Hitherto, even Conservative governments had accorded local authorities a significant and continuing role in housing provision, and local authority house building was accepted as a key instrument of policy responses to the nation's housing problems. By the late 1980s, however, Ministers had gone beyond presenting local authority housing as second best to owner-occupation; they began to depict it as 'a problem rather than a solution'.

Nicholas Ridley, Secretary of State for the Environment at the time of the 1987 general election, put out an election statement which asserted: 'The Housing Policy we have pursued has given us a flourishing owner occupation sector. The Right to Buy has been an unqualified success. People want to be home owners'. But the statement went on to say:

> Monopoly provision by local authorities has not succeeded. There have been serious problems of management. Local authority rent arrears now total over £200 million. Thousands of dwellings are left empty. Performance on repairs is poor and insufficient attention is paid to the wishes of tenants. Many estates are plagued by vandalism. For many there is no escape from these conditions. In many areas tenants have become totally dependent on their council.
>
> (Ridley 1987)

Ministerial speeches were clearly intended to undermine local authority housing, both to justify alternative policies and to divert consumers into other apparently more successful tenures. It is important also to recognize that the attack on local authority housing is part of a wider campaign against local government itself. Whereas central government claims credit for what it sees as the successful parts of its housing policy (such as the right to buy), responsibility for the perceived failings of council housing is placed firmly on the local authorities.

Local authority housing management has been the target of considerable criticism in recent years. It was perhaps inevitable in the context of an increasingly aged stock and a much reduced supply of new housing that attention would shift to the management of the existing stock. The complaints of tenants have been seized upon by a government opposed not just to the way local authority housing is managed but to the very existence of a substantial municipal housing service. Support for the Government's attack on standards of housing management has come from the Audit Commission, which published a report bluntly titled *Managing the Crisis in Council Housing* (Audit Commission 1986a). This report called for a range of improvements in housing management, including a strengthening of the position of the chief officer in relation to the elected representatives, referring to 'political interference' in day-to-day management activities. The Commission claimed that if its proposals were implemented by local authorities then there would be substantial savings and benefits; 'excessive' administration costs could be reduced by up to £100 million per year, the number of empty properties could be reduced by up to 25,000, use of bed and breakfast hotels for the homeless could be reduced by a quarter and rent arrears could be cut by £100 million or more. This severe criticism of current local authority performance was followed up in a further report focusing on housing maintenance (Audit Commission 1986b). The Commission's assessment of the problem was that many authorities had neglected repairs over a long period; had operated very slow and inefficient repairs services; and charged rents which were too low to ensure an adequate standard of maintenance.

This official and semi-official criticism of local authority housing was supplemented by a certain amount of influential research by academics such as Coleman (1985), Power (1987) and Minford, Peel and Ashton (1987). Interestingly, however, the studies of housing management commissioned by the Department of the Environment (1989) and the Welsh Office (1989), did not produce conclusions which showed local authorities to be systematically less effective than housing associations.

The significance of the barrage of criticism of local authority housing in the late 1980s was that it came in the context of the shift of Government attention towards the continuing need for a substantial amount of rented accommodation. The Government's view was that although there was still a need for rented housing the local authorities had shown themselves to be inappropriate landlord agencies, and therefore it was necessary to develop policies to encourage investment in private renting and housing associations. Commercial profit-seeking landlords and housing associations were bracketed together in what the Government sought to label the 'independent rented sector'. In the White Paper of September 1987, the first policy objective was stated to be the continued spread of owner-occupation, and the second was to 'put new life into the independent rented sector' (DOE 1987); the third objective referred to local authorities adopting a reduced role as providers of housing, but an enhanced role as 'enablers'.

There were two distinct strands of policy in the broad strategy of reprivatization (Kemp 1990) and demunicipalization (Kemp 1989) of rented housing. The first was the deregulation of the private sector and the introduction of a new financial regime for housing associations. These measures were designed to generate higher levels of investment and to expand the supply of new non-local authority housing. The second element was to promote the break-up of the local authority stock by various devices which would retain it as rented housing but under new, non-local-authority landlords.

### The 1988 Housing Act

The Housing Act 1988 introduced a form of creeping deregulation of the private rented sector; existing 'regulated tenancies' (subject to 'fair rents') remain unchanged, but all new lettings since 15 January 1989 are either 'assured' or 'assured shorthold' tenancies. The key feature of the new system is that market rentals prevail in new lettings and will eventually spread through the sector as a whole. This gives landlords a higher and more competitive rate of return, and other features of the new tenancy arrangements give landlords easier access to vacant possession. From the tenants' perspective this obviously means that they face higher rents with less security of tenure. However, the Government's position would be that tenants would not be disadvantaged if the new system generated a significantly increased supply of housing. It is still too early to make a definitive judgement, but Whitehead and Kleinman (1989, p. 77) argue that a number of factors, not the least of which is the financially privileged position of the owner-occupier, mean that a significant revival of private renting is unlikely:

> Under current conditions . . . the evidence suggests that the long run equilibrium size of the sector would be significantly less than its current level. Unless therefore there are increases in effective demand or reductions in costs it is unlikely that the continuing erosion of the traditional private rented sector will even be fully offset by new investment generated as a result of the changing regime.

In the housing association sector, the 1988 Act introduced assured tenancies for all new lettings and required associations to set their own rents (after fifteen years of rents being set by the rent officer service). The funding of housing association development was also fundamentally altered, in a way designed to reduce the overall level of subsidy and public sector investment and to increase the amount of private finance (Cope 1990). This can be seen as a form of reprivatization since it pushes associations back towards their pre-1974 position and requires them to adopt a much more commercial approach. The new system is based on higher rents and lower general subsidy, supported by housing benefit targeted on tenants with the lowest income. This may lead to higher levels of investment, but by mid-1990 the Housing Corporation was in a financial crisis as associations raised the pace of development (Social Housing 1990; Pollitt 1990).

The second strand of policy in the 1988 Act consisted of two elements: 'tenants' choice' and HATs. Labouring under the belief that local authority tenants were widely experiencing feelings of being trapped in the public sector, the Government introduced 'tenants' choice' as a way of facilitating the transference of estates into non-municipal ownership. Although presented as a chance, even a right, for tenants to opt out of local authority control, the drafting of the Act in fact established a right for private landlords to buy in to the stock, although tenants have a right to vote against any proposed takeover. (Any tenants, however, who do not vote *against* a transfer are deemed to have voted in favour!) During the passage of the Bill through Parliament there was widespread fear on local authority estates that predatory landlords would be able to pick off the most desirable or most profitable parts of the stock. In practice, however, in the first eighteen months after the Act took effect the impact has been negligible. Interest from private landlords has been almost entirely absent, and in mid-1990 the approved list of landlords contained only one which was not a registered housing association (Roof May–June 1990). Transfer activity has been concentrated in areas where the local authority itself has decided to dispose of its entire stock to a new or existing housing association; by March 1990 ten authorities had successfully taken this route (Roof March–April 1990).

The Government's other attempt in the 1988 Act to break up the public sector was the declaration of Housing Action Trusts (HATs). Under Part III of the Act, the Secretary of State for the Environment can declare HATs in areas of mainly local authority housing which are deemed to need injections of finance and management expertise beyond the ability of the council itself to provide. Here too the

Government appears to have misread the mood of tenants because reaction in the first round of six HAT declarations was generally strongly opposed to the proposed separation from local authority ownership. HATs now appear to have a very limited future.

However, the 1988 Act was followed by the Local Government and Housing Act 1989, which may yet have the effect of triggering a larger-scale break up of public housing and a revival of interest in tenants' choice and other transfers. Space does not permit a full discussion of the Act (see Malpass and Murie 1990; Malpass 1990, 1992) but there are two main points to be made. The Act introduces new arrangements for local authority capital expenditure and a new subsidy system for local authority housing. As a result of the sale of houses and other assets during the 1980s, local authorities accumulated huge capital receipts (total capital receipts in 1979–89 amounted to £17,580 million (Malpass and Murie 1990, p. 92). The gradual disposal of these receipts, within the rules of the 1980 system, meant that central government had less control than it wanted over the scale and distribution of local authority capital spending. The 1990 system gives the centre much tighter control of local authority capital programmes, and is widely expected to lead to significant reductions in resources available for maintaining the quality of the public sector housing stock. A report by the Association of District Councils forecast a fall of 27 per cent in its members' capital spending on housing in 1990–1, and a further fall of 36 per cent in 1991–2 (ADC 1990, p. 21). This in itself may be enough to trigger much wider interest in voluntary transfer of local authority stocks to new owners who would not be subject to the same constraints.

At the same time, the new subsidy system introduced in Part VI of the 1989 Act permits the Secretary of State to exert very considerable pressure on local authority rents and this, too, may be a powerful incentive to tenants to seek alternative landlords. The new system is similar to the 1980 system, in so far as it provides a subsidy to make good any deficit on the notional housing revenue account (HRA), irrespective of the state of the actual account. However, there are three crucial differences between the old and the new system. First, the HRA is now 'ring fenced', which means that authorities cannot pay discretionary subsidies into the account, nor can they transfer surpluses out of the account. Second, there is now one subsidy from central government, known as the HRA subsidy, and this includes what was previously the separate rent rebate subsidy. The key to the centre's leverage on rents lies in the rolling together of general and means-tested subsidy into one deficit subsidy.

This is effectively a way of ensuring that as rent income rises relative to expenditure the benefits will accrue to central rather than local government. Third, the new system involves an attempt to differentiate guideline rent increases and management and maintenance allowances according to local circumstances. Thus in April 1990 rents were deemed to have increased by amounts varying between 95p and £54.50 per week, according to the relationship between existing rents and local capital values. In fact, rent increases varied between zero and £16.75 per week. The overall average guideline increase set by the DOE was 10 per cent but the ADC survey reported an actual average of 16 per cent (ADC 1990, p. 7) and figures published in Roof (July–August 1990, p. 16) show that the top twenty percentage increases were all 30 per cent or more.

This section has reviewed the period since 1979, highlighting the continuing validity of Donnison and Maclennan's (1985) view that it was a time of tenure policy rather than housing policy. Traditional housing policy concerns have been supplanted by the pursuit of tenure change. The Conservatives under Margaret Thatcher have proudly proclaimed themselves as the Party of the home-owner, and in pursuing that policy they have gone further than any previous government in attacking and undermining local authority housing.

That attack has consisted of four main elements: break-up of the stock by sales to individuals and transfers to alternative landlords; cuts in resources available for new building and, from April 1990, capitalized repairs; increases in the power of central government to drive up rents and reduce subsidy; and a campaign by Ministers and their supporters to denigrate local authority performance in housing provision and management. The emphasis on legislative intervention in relation to local authority housing contrasts markedly with the refusal to tackle the hugely expensive issue of tax relief for mortgaged owner-occupiers and the absence of any real taxation of the wealth accumulated through home-ownership. During the 1980s the cost of mortgage interest tax relief rose from £1,600 million in 1979–80 to £7,000 million in 1989–90. In real terms tax relief more than doubled in this period, reflecting a considerable growth in the numbers of mortgaged owners.

The Thatcher Governments have also displayed a belief in the ability of private renting to revive, but as a kind of insurance policy they have espoused a reformed and privatized housing association sector as the mainstay of rented housing provision in the 1990s. The Tories have shown little concern for planning total housing output, and for most of the decade it seemed that the Government was unconcerned about the plight of the homeless. There are grave contradictions in the way that different arms of Government policy affect housing. For instance, there was in the early 1980s a conflict between the housing policy objective of expanding owner-occupation and the economic policy which raised both interest rates and unemployment. In the case of rented housing the policy has been on the one hand to encourage local authorities and housing associations to concentrate on the least well off, yet rents have been forced up and the housing benefit system has been made less and less generous. The clear implication of all this is that policy has contributed to a gathering affordability crisis, but what is the evidence?

## THE AFFORDABILITY CRISIS

The term affordability has been in use for some years in relation to housing provision in the Third World, but in Britain it has been in common usage only since the late 1980s. There were two main factors which led to considerable public debate about affordability in British housing. The first was the White Paper on housing in September 1987 (*Housing: The Government's Proposals*, CM 214) which proposed changes in the subsidy systems for social rented housing and an end to the fair rent system, which had been the basis of rent setting in housing associations for fifteen years. The second factor was the boom in house prices in 1988–9, followed by rising interest rates and falling prices.

The proposal to end the fair rent system for new lettings implied free market rents in the private sector, raising questions about whether this would lead to a real increase in supply or merely to higher rents, and thus to an effective decrease in the availability of accommodation within the reach of people on low incomes. In the case of housing associations, the policy was not market rents but 'affordable rents' set by the associations themselves, rather than by the independent rent officer service. This raised the issue of what constituted an affordable rent, and whether the new policy would ultimately result in a change in the social composition of housing association tenants. The White Paper itself referred to housing association rents 'at levels which are within the reach of the tenants for whom they traditionally provide', (paragraph 4.7) but there was no more specific indication of the levels implied nor of the criteria for rent setting. A separate consultation paper was similarly vague, referring only to the way in which the new grant system would ensure that associations could set 'sensible rents' attuned to the means of prospective tenants. Government Ministers and the Housing Corporation continued to refuse to define what affordability might mean in the housing association context, and in the absence of any clear official guidance the National Federation of Housing Associations produced a case for taking 20 per cent of the median net income of new tenant households in work as a rent guideline.

This has subsequently been adopted as a benchmark, but it is open to criticism on a number of grounds. First, the 20 per cent figure can be seen as an arbitrary amount, without any coherent justification, although it can be defended as being rather less than the proportion spent by first-time buyers (Bayley 1988). Second, rent to income ratios refer to averages, and some people inevitably find themselves paying substantially higher proportions of their incomes on rent. This sort of problem can, in theory, be overcome via income-related assistance, but in Britain the housing benefit system was modified in the 1980s to make it less helpful in this context. Wilcox (1990) has referred to what he calls the 'affordability mountain' which reflects the way in which the steep taper applied to housing benefit causes the proportion of income devoted to rent to rise well above 20 per cent as income rises, although it then tails off again. A more generous taper would be required in order to protect tenants from both very high marginal tax rates and rents well above 20 per cent. Wilcox (p. 52) concludes that:

> The problem of affordability defined in this way is less to do with higher rent levels, and more to do with the inadequacies of the housing benefit system. Improvements to the housing benefit system, and in particular a substantial reduction in the 65% rent taper, are a pre-requisite for any coherent policy towards higher rents for social housing.

A third objection to defining affordability in terms of crude percentages is that there is an important element of subjectivity involved in what consumers themselves are prepared to pay for housing. Affordability ultimately depends upon the proportion of income that consumers are willing and able to devote to housing. The three crucial variables here are income, the non-housing demands on that income and the consumer's perception of value for money. In other words, the important determinant of what consumers regard as affordable is the scope for trade-offs between different forms of expenditure and their relative attractions. Thus households with very high incomes might decide to allocate a high proportion to housing for locational or status reasons, secure in the knowledge that such a decision left sufficient resources for non-housing expenditure. On the other hand, low-income households might find that devoting a similar proportion of income to housing left an inadequate cash balance to meet non-housing needs. Household size and composition also influence the trade-off decisions that people make, but what they are prepared to spend on housing and hence what counts as affordable, must reflect, at least to some extent, individual perceptions of what they are getting for their money. Thus it can be argued that newly-mortgaged owners in Britain are typically prepared to allocate a higher proportion of income to housing than are tenants because of the investment element in mortgage costs, and because owners normally expect housing costs to fall in real terms during the loan repayment period.

However, the subjectivity of affordability is constrained by market conditions and by factors such as the availability of loan finance. Mortgage lenders normally limit loans to a standard multiple of earnings, thereby putting a ceiling on the proportion of income devoted to house purchase. Unfortunately this ceiling can rise if mortgage rates go up, or if income falls, and then the home-owner can be faced with difficult choices.

There are grounds for saying, then, that affordability is a virtually undefinable concept and certainly cannot be neatly or simply understood in terms of a fixed percentage of income. The government and the Housing Corporation have staunchly refused to become closely involved in attempts to resolve the debate, and the Director General of the Building Societies Association was quoted in 1990 as saying, 'The affordability debate . . . is one of the most sterile I have ever heard'. (Grant 1990). Nevertheless, there is evidence of a growing problem of affordability in both rented and owner-occupied housing in Britain in the late 1980s and early 1990s.

### Homelessness

The starkest evidence of the contemporary housing crisis is the growing incidence of homelessness, which is at its most visible in the centres of Britain's big cities where the numbers of people sleeping in doorways and begging on the streets have increased markedly in recent years. The so-called 'cardboard city' of homeless people on London's south bank has become a powerful touchstone for those campaigning on behalf of the homeless, although estimates of the numbers involved are inevitably open to dispute. One estimate of the scale of the problem in London in 1986-7 was that up to 3,000 people were sleeping rough (quoted in Greve and Currie 1990, p. 10), and the numbers are known to have increased substantially since then.

Official homelessness in England, measured in terms of the households accepted as homeless by local authorities, doubled between 1978 and 1989, from 53,100 to 126,680 households. The total number of households claiming to be homeless in 1989 was double the number accepted (Oldman 1990, p. 3). Figures for 1990 indicate a continuing escalation of official homelessness, with acceptances in the third quarter running 14 per cent ahead of 1989 (*Inside Housing* 14 December 1990). In September 1990 there were nearly 46,000 homeless households placed in temporary accommodation, compared with 32,000 in 1989. An important element in the increase in homelessness is the rising number of households becoming homeless because of mortgage arrears. The numbers are still not large in the overall context of homelessness, but there was a 50 per cent increase in the year to September 1990 (*Inside Housing* 14 December 1990).

Households accepted as homeless represent only a small proportion of all those who are on local authority waiting lists. In 1989 there were 1.25 million households on council lists (Oldman 1990, p. 13), although there is considerable doubt about the value of waiting lists as a true indicator of housing need. Nevertheless it is clear that the numbers rehoused from waiting lists have fallen in recent years, and that the proportion of lettings taken by homeless households has increased; in 1987-8 31 per cent of all new tenants in England and Wales were housed as homeless (National Housing Forum 1989, p. 13).

Perhaps the most influential recent work on the estimation of housing need is the series of studies carried out by Glen Bramley (1987, 1988, 1989 and 1990). In the earlier of these studies Bramley looked at not only the total need for new social rented housing but also the geographical distribution of need. The conclusion that the need was greatest in the south of England generated some criticism from housing interests in the north (Barnett, Bradshaw and Lowe 1989), but as Niner has commented, 'Critics do not . . . dispute the general case that there is a considerable unmet demand for council housing'. (National Housing Forum 1989, p. 15).

In later work, Bramley (1990) sought to develop the technique for measuring affordability and estimating the number of households who could not afford access to owner-occupation. In principle the task is straightforward: the price of new houses in each area is compared with income data to show how many households can and cannot afford to buy, given certain assumptions about mortgage borrowing. Bramley worked on the basis of a 100 per cent mortgage of three times annual income for single-earner households and two and a half times joint income for two-earner households. His research showed that in all regions of England, including the cheapest, there were substantial proportions of households who, in terms of their ability to raise a mortgage, could not afford even a one-bedroom starter home, and larger proportions in each region who could not afford a three-bedroom family house. He notes that,

> It is salutary to reflect that the most characteristic product of the contemporary private housebuilding industry, a family house in South East England, could not be afforded by 75% of households living in that region on the basis of current income alone, disregarding assets.
>
> (Bramley 1990, p. 16)

Bramley's work highlights the severity of the affordability problem in the south-east of England, and suggests that, even if low cost and shared ownership options are developed, there will still be virtually a third of new households, taking the country as a whole, who need rented housing. The continuing need for rented housing is confirmed by a report from the London Research Centre (1990) which looked at the relationship between prices and incomes in the capital:

> over half of full-time workers in London (in 1990) would have been unable to purchase even the very cheapest one bedroom flat on their own. And even doubling up in these cheapest one bedroom (and probably studio) flats would still leave a significant proportion of full-time workers absolutely excluded.

### Owner-occupiers' problems

From the point of view of individual households wishing to purchase in the owner-occupier market there are three relevant indicators of affordability: i) Are prices of suitable houses in appropriate locations such that mortgage lenders will finance the initial purchase at the prevailing level of income? ii) Is income sufficient to maintain mortgage repayments if interest rates rise or domestic circumstances change? iii) Given the burden of mortgage costs, is there sufficient income and/or credit available to maintain the physical fabric of the dwelling?

House prices generally fluctuate in a cyclical pattern as the market moves from boom to slump, and as Ball (1986) has argued, instability has become a chronic feature of the British housing market since the early 1970s. For the individual purchaser the cost of housing relative to income depends to some extent on the point in the cycle at which the purchase is made. The difficulties being experienced by owner-occupiers in the early 1990s result, in many cases, from a highly unfortunate sequence of events in 1988 to 1990. People who bought houses in the summer of 1988, when prices were approaching a peak, often did so by borrowing heavily and at a time when interest rates were at their lowest point for a decade. The market was artificially stimulated by a decision by the Chancellor, announced in his Spring 1988 budget, to end the system of 'double tax relief' (whereby unmarried couples could both qualify for relief on loans up to £30,000). In the period up to the August deadline there was, therefore, hectic activity in the housing market, and almost inevitably this was followed by a lull which became a slump as interest rates rose rapidly. Mortgage interest rates, which had stood briefly at 9.5 per cent in April 1988 climbed steadily to 15.4 per cent (an all time record) by February 1990 as a direct result of the Government attempting to use housing costs as a device to dampen consumer demand in the economy as a whole.

Figures published by the Council of Mortgage Lenders (1990, Tables 7 & 8) show that in 1988 and 1989 the ratio of average loan to average earnings stood at a level which, for first-time buyers, had not been reached since 1973, and in the case of 'selling buyers' the ratio was at an all-time high. The steep increase in interest rates meant that repayments which had been just affordable at the point of purchase became very much heavier. The cost of the average first-time buyer loan of £30,000 in 1988 rose by £110 per month by early 1990. In London and the South East first-time buyers in 1988 were generally having to borrow far more than £30,000. The cost of a £50,000 loan would have increased by no less than £308 per month between April 1988 and February 1990. These sorts of increases were far too large to be absorbed by the effect of inflation on net earnings, and also too large to be dealt with by spending cuts elsewhere in household budgets.

In these circumstances there was inevitably a big increase in the number of households who got into difficulties with their mortgage repayments, and harrowing accounts of innocent, respectable families engulfed by housing debt appeared in the newspapers and current affairs programmes. In popular imagery they ranked with cardboard city and young people sleeping in doorways as indicators of the housing crisis in the early 1990s.

In 1988 and 1989 building society repossessions actually fell, but in 1990 the total rose to nearly 44,000, moe than twice the 1988 peak and an all time record (*The Guardian* 15 February 1991). In the context of the overall total of outstanding mortgages these numbers remain very small, but the rate of increase is dramatic, indicating the severity of market conditions in 1990. It must be remembered, also, that the figures quoted here refer to building society activity only and do not include action taken by other lenders.

Turning to mortgage arrears as another indicator of affordability problems, the official building societies' statistics show a 69 per cent increase in arrears of 6–12 months in the first half of 1990 compared with the same period in 1989, from 45,100 to 76,280. Arrears of over 12 months rose even more sharply, by 102 per cent, to 18,750. However, the building society arrears figures always exclude people owing less than six months' repayments. Ford (1990) carried out a survey in April 1990 which suggested that as many as 443,000 mortgages were in arrears of 1–5 months, and that overall the proportion of mortgages in difficulty was nearly 6 per cent, with much higher proportions amongst people who had recently taken out their loans.

Owners struggling with unaffordable mortgage repayments, inflated by interest rate increases, cannot always escape their difficulties by selling their houses and moving to somewhere cheaper. People who bought when the market was at its most active in 1988 have generally seen their houses decline in value, and those with very high percentage loans may have come to owe more than their houses are worth. The Nationwide Anglia Building Society reported that in the year to December 1990 there was a 10.7 per cent fall in house prices, the first time this had been recorded since 1954. In real terms, average prices fell by 14 or 15 per cent in the year to December 1990.

### Rent levels

Turning to rent levels, a report by the National Association of Citizens' Advice Bureaux (NACAB 1990) showed that private sector rents rose sharply after deregulation in 1988, and that many tenants coming for advice were experiencing difficulty in paying higher rents. The report concluded that the private rented sector was becoming increasingly inaccessible to low-income households. However, the fair rent system continues for tenants who remain in occupation, but here too there is evidence (Critchley 1990) that rent officers have been increasing fair rents at a rate running far ahead of inflation.

Housing association data (NFHA 1990) show that rents, both fair and assured, have increased rapidly in real terms since the start of 1989. In the period from the second quarter of 1988 to the first quarter of 1990 housing association rents rose by 43.6 per cent, while the retail prices index rose only 12.7 per cent, and the index of tenants' incomes rose only 11.2 per cent. In terms of the 20 per cent affordability ratio, assured tenancy rents were running at 24 per cent of median net weekly incomes by March 1990.

At the same time, council rents were also rising in real terms, although in cash values they generally remained lower than housing association rents. The average council rent in England and Wales in April 1990 was £23.42 per week, an increase of 13.5 per cent on the previous year, but, as mentioned earlier, some councils increased rents by well over 30 per cent in response to the new financial regime for local authority housing.

## CONCLUSIONS

This chapter has approached the housing question in Britain by arguing that there is a serious affordability problem, affecting all tenures, and that its severity has been exacerbated, if not directly caused, by recent and current government policies. In particular, the argument has been developed that issues of housing tenure have been given undue emphasis in housing policy, at the expense of more conventional concerns with quality, quantity and affordability. In a situation where the poorer sections of the population are generally renting their homes and the better off are overwhelmingly home-owners, it is perverse to adopt policies which are designed to raise rents while defending the hugely expensive and highly inequitable system of mortgage interest relief. Nevertheless, the obsession with promoting individual home-ownership and the fantasy of a thriving private rented sector have led to these policies being imposed, and retained despite mounting evidence of problems throughout the housing system. During the 1980s the Government's ideological commitment to home-ownership outstripped the capacity of the housing system. Consumers were directed towards home-ownership to an extent unjustified by incomes and economic circumstances. As a result the pain inflicted on marginal and over-extended mortgage borrowers by the instability of the market was much worse, both qualitatively and quantitatively, than it would have been in the context of a more balanced approach to meeting housing needs. The recession in the housing market in the late 1980s and early 1990s coincided with the failure of the private rented sector to respond as hoped, the continuing enfeeblement of the local authority sector and the crisis of development funding for housing associations.

In thinking about how to respond to Britain's housing problems it is important to take into account the connections between housing and the wider economy. This is a theme which emerged with some force by the end of the 1980s, and was starkly illustrated in the way that in 1988–9 the Government used rising interest rates to slow down both the housing market and the economy as a whole. Rising mortgage costs were deliberately engineered in order to reduce consumer spending in general. Financial liberalization in the early 1980s helped to stoke up house prices by 1988, and facilitated equity withdrawals at levels far higher than would have been possible in earlier years. The growth of home-ownership, accompanied by rapid house price inflation, has resulted in a situation where the total value of housing assets held by individuals exceeds aggregate mortgage debt by some £750 billion (Spencer 1990). This represents a huge potential stimulus to further bouts of consumer spending, especially in a context where people are increasingly encouraged to cash in their assets to pay for short-term consumer spending. Estimates of the amount of equity leaking out of housing into wider spending vary, but Spencer (1990) claims that over £11 billion was withdrawn in 1988. There is, therefore, a conflict between the political need to reduce mortgage interest rates in order to ease the pressure on hard-pressed home-owners, and the need to avoid a consumer spending boom when rates fall.

Recognizing the links between housing and the wider economy is important for raising the political profile of housing, but at the same time it tends to make the development of reform policies that much more difficult. The need to reform housing finance, for instance, cannot be satisfied by proceeding within narrow housing boundaries, but must be situated in relation to broader economic considerations.

Nevertheless, there are some specific housing measures which could be taken to tackle the problems faced by households in the worst housing difficulties. Some of these measures have been referred to already in this chapter. For instance, a revival of investment in new building by local authorities is necessary to tackle homelessness in a more robust way than that implied in recent high-profile but short-term packages. Local authorities, however, should not be expected to produce large numbers of new dwellings at break-neck speed – lessons have to be learned from past mistakes. A programme of 50,000 completions per year within three years is achievable and sustainable, without resort to short cuts and quick fixes. The housing associations clearly have a big role to play in meeting the continuing need for social rented housing, but the local authorities still have great advantages as providers of affordable accommodation. Their existing asset base and ability to spread costs by rent pooling make them essential leaders in the field.

Local authorities also have a role to play in supporting home-ownership. Forrest, Murie and Williams (1990) refer to the idea of developing services which combine the best features of different tenures, including 'an alternative home-ownership package supported by public subsidy directed to achieve clear social objectives' (p. 210). They go on to say:

> It may be argued that such arrangements could be developed to enable local authorities to redirect their attention to investment and repair, intervention in access/exchange, provision of aid and advice and using the full range of their resources to deliver housing to households rather than day-to-day management in the model of the private landlord.

Ideas of this sort deserve careful consideration in an era of mass home-ownership.

The reform of housing finance has been debated in Britain for twenty years without any government having the will or the courage to take on the task in a coherent manner. Indeed, most have adopted policies which have been driven by the desire to promote home-ownership rather than the rational reform of housing finance, and as a result the financial system has become ever more distorted. It is not appropriate to set out here a detailed programme for financial reform in housing, but it has to be said that restructuring of mortgage interest relief must be a priority for any attempt to create a fairer and more equitable system. Some method for phasing out interest relief after eight or ten years seems to be the best approach, together with a restriction on relief so that it is confined to the amount of the initial (rather than subsequent) loan.

In relation to rents there is a well-represented body of opinion in favour of rents which reflect the real cost of provision, so that there is a clear basis for investment decisions and overall resource allocation. However, for such an approach to work fairly it is necessary to have a housing benefit system which is much more generous than the present British arrangements. In specific terms this means having a shallower taper. The reform of housing benefit should be seen as a top priority in any attempt to overcome the affordability crisis. More

generally it is desirable to move towards a reduction in the distortion
of tenure choice which has developed into a chronic condition of
housing in Britain.

## REFERENCES

Allen, J. & MacDowell, L. (1988) *Landlords and Property*, Cambridge:
  Cambridge University Press.
Association of District Councils (1990) *Survey on Council House Rents, Housing
  Subsidy and Capital Expenditure*, London: Association of District Councils.
Audit Commission (1986a) *Managing the Crisis in Council Housing*, London:
  HMSO.
—— (1986b) *Improving Council House Maintenance*, London: HMSO.
Ball, M. (1983) *Housing Policy and Economic Power*, London: Methuen.
—— (1986) *Home Ownership: A Suitable Case for Reform*, London: Shelter.
—— (1988) *Rebuilding Construction*, London: Routledge.
Barnett, R., Bradshaw, J. & Lowe, S. (1989) *Not Meeting Housing Needs, a
  critique of the ADC Report 'Meeting Housing Needs'*, York: University of York.
Bayley, R. (1988) 'The Rent Dilemma', *Housing*, October, and 'Fair is not
  Affordable', *Housing*, December.
Bentham, G. (1986) 'Socio-Tenurial Polarisation in the United Kingdom,
  1953–1983: the income evidence', *Urban Studies*, 23 (2): 157–62.
Boddy, M. (1989) 'Financial Deregulation and UK Housing Finance',
  *Housing Studies*, 4 (1): 92–104.
Booth, P. & Crook, T. (eds) (1985) *Low Cost Home Ownership*, Aldershot:
  Gower.
Bramley, G. & Paice, D. (1987) *Housing Needs in Non-Metropolitan Areas*,
  London: ADC.
Bramley, G. (1988) *Access to Owner-Occupation: Research Note*, London:
  ADC.
—— (1989) *Meeting Housing Needs*, London: ADC.
—— (1990) *Bridging the Affordability Gap*, Birmingham: BEC Publications.
British Market Research Bureau (1986) *Housing and Savings 1986*, Technical
  Report and Tables, London.
Building Societies' Association (1983) *Housing Tenure*, London, BSA.
Coleman, A. (1985) *Utopia on Trial*, London: Hilary Shipman.
Cope, H. (1990) *Housing Associations: Policy and Practice*, Basingstoke:
  Macmillan.
Council of Mortgage Lenders (1990) *Housing Finance*, No. 7 August.
Critchley, R. (1990) 'Scarcely Credible', *Roof*, November-December.
DOE (1987) *Housing: the Government's Proposals*, London: HMSO, Cm 214.
—— (1989) *The Nature and Effectiveness of Housing Management in
  England*, London: HMSO.
Donnison, D. & Maclennan, D. (1985) 'What Should We Do About
  Housing?' *New Society*, 11 April.
Ford, J. (1990), 'National Debts', *Roof*, September–October.
Forrest, R. & Murie, A. (1988) *Selling the Welfare State*, London: Routledge.
Forrest, R., Lansley, S. & Murie, A. (1984) *One Foot on the Ladder*, Bristol:
  School for Advanced Urban Studies.

Forrest, R., Murie, A. & Williams, P. (1990) *Home Ownership: Differentiation and Fragmentation*, London: Unwin Hyman.

Grant, C. (1990) 'Mark My Words', *Roof*, March–April.

Gray, F. (1979) Chapter 8 in Merrett, S. *State Housing in Britain*, London: Routledge & Kegan Paul.

Greve, J. & Currie, E. (1990) *Homelessness in Britain*, York: Joseph Rowntree Memorial Trust.

Hamnett, C. (1984) 'Housing the Two Nations: Socio-Tenurial Polarisation in England and Wales, 1961–81' *Urban Studies*, 21 (2): 389–405.

Hills, J. (forthcoming) *Unravelling Housing Finance*, Oxford: Oxford University Press.

Institute of Housing, (weekly) *Inside Housing*, London.

Kemp, P. (1989) The Demunicipalisation of Rented Housing, in Brenton, M. & Ungerson, C. (eds) *Social Policy Review 1988–89*, London: Longman.

—— (1990) 'Shifting the Balance Between State and Market: the reprivatisation of rental housing provision in Britain', *Environment and planning A*, 22: 793–810.

London Research Centre (1988) *London Housing Survey 1986–87*, London: LRC.

—— (1990) *House Prices in London*, Quarterly Bulletin 15.

Malpass, P. (1990) *Reshaping Housing Policy*, London: Routledge.

—— (1992) *Implementing Housing Policy*, London: Routledge.

Malpass, P. & Murie, A. (1990) *Housing Policy and Practice*, 3rd edition, Basingstoke: Macmillan.

Minford, P., Peel, M. & Ashton, P. (1987) *The Housing Morass: Regulation, Immobility and Unemployment*, London: Institute of Economic Affairs.

NACAB, (1990) *Market Failure: low income households and the private rented sector*, London.

National Housing Forum (1989) *Housing Needs in the 1990s: an interim assessment*, London.

NFHA (1990) *Core Quarterly Bulletin No 2*, London.

Oldman, J. (1990) *Who Says There's No Housing Problem?* London: Shelter.

Pickup, D. (1986) *When We Build Again: Partnership*, unpublished paper presented at a conference, 'When We Build Again', SAUS, University of Bristol.

Pollitt, N. (1990) 'Grave New World', *Roof*, July–August.

Power, A. (1987) *Property Before People*, London: Allen & Unwin.

Ridley, N. (1987) Press release from Conservative Central Office, 19 May.

Robinson, R. & O'Sullivan, A. (1983) 'Housing Tenure Polarisation, Some Empirical Evidence', *Housing Review*, July–August.

*Social Housing* (1990) 'Crisis? What Crisis?' May.

Spencer, P. (1990) 'The Credit Monster Leaves Home', *Roof*, May–June.

Treasury (1989) *The Government's Expenditure Plans 1989–90 to 1991–92*, London: HMSO.

Whitehead, C.M. and Kleinman, N. (1986) *Private Rented Housing in the 1980s and 1990s*, Cambridge.

Wilcox, S. (1990) *Social Housing in the 1990s: Challenges, Choices and Change*, London: Institute of Housing.

Willmott, P. & Murie, A. (1988) *Polarisation and Social Housing*, London: Policy Studies Institute.

# 4 Housing affordability in the Federal Republic of Germany

*Rudi Ulbrich and Uwe Wullkopf*

In the first two decades of its existence, the Federal Republic of Germany experienced a 'housing miracle'. After the Second World War, the country was in ruins; half the housing in the large cities had been destroyed or rendered uninhabitable and millions of refugees from the East were pouring into the country. And yet by the early 1970s West Germans, at all levels in society, were better housed than their predecessors had ever been. However, new problems of affordability were beginning to emerge. The 1980s saw a dissolution of the policy mix which had been adopted at the birth of the Federal Republic. More market-orientated policies were adopted, which emphasized housing allowances and tax incentives for owner-occupation rather than the provision of social housing. This neo-liberal policy culminated with the abolition in 1988 of the non-profit housing movement. To reach a verdict on the policies that have been pursued in the 1980s – and so to be in a position to frame appropriate housing policies for the 'new' Federal German Republic, it is necessary to first review the policies and achievements of the post-war period.

## THE POST-WAR GERMAN 'HOUSING MIRACLE'

The housing policy which underlay the post-war 'housing miracle' has to be seen in the context of the general economic and political institutions of the new state. The catastrophic situation in 'year zero', and the experience of National Socialism, led to a profound political change, and a return to older German political ideals. The political system of the new German Federal Republic was – and, as the new united Germany, has remained – decentralized, with extensive functions, and taxing powers devolved to regional (*Laender*) and local governments. Thus the *Laender* undertake the administration of housing policy and have some limited powers to modify federal housing legislation; housing subsidies are provided by both the Federal and the *Laender*. Local authorities also play an active 'enabling role' in the housing market, and are free to subsidize housing activities out of their own revenues. They do not normally own housing directly, but in the larger cities there are housing associations (formerly non-profit) in which the municipality is the principal shareholder.

The dominant economic philosophy of the Federal Republic (first expounded by Christian Democrats but later accepted and developed by Social Democrats) was *soziale Marktwirtschaft* ('a socially responsible market economy') which combined both market and social elements. In the field of housing, this meant that the market system has been encouraged in both owner-occupation and private tenancy, but that social objectives have also been pursued through (especially in the early days) the provision of social housing and, from the 1960s onward, through housing allowances (*Wohngeld*).

The electoral system of proportional representation, and popular sentiment, have ensured that all Federal Governments have been coalitions; first of Christian Democrats and Free Democrats; then of all three parties; then of Social Democrats and Free Democrats; and since 1981 of Christian Democrats and Free Democrats once again. The outcome has been more continuity of policy than in many other countries, in spite of the trend towards economic neo-liberalism in the 1980s.

**Post-war policies**

The housing policy adopted in the early years of the Federal Republic by Right-Liberal Governments was eclectic. It sought to utilize all possible resources for the great task of reconstruction by encouraging a vigorous 'market' sector (especially in private tenancy) as well as a 'social' housing sector covering both non-profit associations and private landlords. Thus the rents of existing housing were at first controlled, but they were periodically raised to preserve their real value. Rents of new, privately financed, rented housing were not controlled, and landlords were given generous depreciation allowances. In the 1960s, rent controls were abolished, and replaced by a provision that rents should be set at a level comparable to that in the municipality (*Vergleichsmieten*). The outcome has been a viable private rented sector, catering for all income groups – although there has been a recent tendency to move up-market. The sector has expanded in absolute size since the 1950s, although it is falling in percentage terms as owner-occupation increases (Table 4.1).

*Table 4.1* Housing stock in the FRG
(millions)

| | Rented | Occupied dwellings Owner-occupied | Total | Empty dwellings | Total |
|---|---|---|---|---|---|
| | | | Subsidized | | |
| 1965 | 3.21 | 1.13 | 4.34 | – | – |
| 1972 | 3.75 | 1.34 | 5.09 | – | – |
| 1978 | 4.07 | 0.95 | 5.02 | 0.08 | 5.10 |
| 1987 | (4.12) | (1.15) | (5.27) | – | – |
| | | | Unsubsidized | | |
| 1965 | 8.90 | 5.21 | 14.11 | – | – |
| 1972 | 9.85 | 6.12 | 15.97 | – | – |
| 1978 | 10.45 | 7.61 | 18.06 | 0.68 | 18.74 |
| 1987 | (11.78) | (9.06) | (20.84) | – | – |
| | | | Total | | |
| 1950 | 6.57 | 4.09 | 10.66 | 0.02 | 10.68 |
| 1956 | 9.13 | 4.80 | 13.93 | 0.08 | 14.01 |
| 1965 | 12.11 | 6.34 | 18.45 | – | – |
| 1972 | 13.60 | 7.46 | 21.06 | 0.33 | 21.39 |
| 1978 | 14.52 | 8.61 | 23.13 | 0.71 | 23.84 |
| 1982 | 14.45 | 9.55 | 24.00 | – | – |
| 1987 | (15.92) | (10.20) | (26.12) | 0.48 | 26.60 |

( ) Estimated
*Source:* Statistisches Bundesamt, – *Wohnungszaehlungen von* 13 Sept. 1950:
*Wohnungsstatistik* 1956/7: *1% Wohnungsstichprobe* 1965 & 1972:
*Ergaenzungserhebung zum Mikrozensus* 1982: *Gebaeude – und Wohnungszaehlung* 25
May 1987

## Social housing

The objective in the early years was not only to encourage privately
financed housing construction but also to produce a large volume of
subsidized 'social housing'. This mainly consisted of rented housing,
although some subsidies were available to low-income owner-occupiers
(the 'second subsidy way'). The subsidies for rented social housing
were made available to both non-profit associations (which had a
history going back to the late nineteenth century) and to private
landlords, provided that they agreed to operate the housing as 'social
housing' for, originally, at least sixty years. This involved letting to
tenants with incomes below the qualifying level, and at the prescribed
rents. The rents were set so as to give the landlord a 'normal'
return on his own capital. The rationale was that the private landlords
were able to contribute capital and management resources which
would not otherwise have been available for social housing. In the
second half of the 1950s, up to 40 per cent of all social housing was

built by private landlords, although the percentage later dropped sharply.

Subsidies for social housing were not given to local authorities as such, but some of the non-profit associations were owned by local authorities, who therefore enjoyed considerable influence, with the right to nominate tenants. Other non-profit associations, such as those sponsored by churches and trade unions, were free to select their own tenants from among those qualifying for social housing.

### The non-profit housing associations

The non-profit associations can be further divided into limited companies and co-operatives. The limited companies built to let to any 'social' tenants, and often also built for sale as well. Under the non-profit legislation, they were subject to various limitations, and in return were freed from taxation. In particular, their activities had to be 'cost covering', and this applied to individual dwellings rather than to the enterprise as a whole. The co-operatives, which own about 1 million dwellings, are locally based, and let only to their own members. The co-operatives have worked well, but their members are mainly middle-income people, with some capital. They have therefore made a contribution to the housing needs of some households which could not quite afford to be owners, but they have not had any significant impact on the provision of housing for the poor.

### The post-war record

Once the housing programme got under way in the early 1950s, it quickly rose to over 500,000 units a year, although output peaked at 670,000 units in 1974. A huge volume of housing, catering for all sections of the population, was built and, by the 1970s, the overall shortage resulting from the War and its aftermath had been ended. After 1974, the number of housing completions declined sharply, to between 300,000 and 400,000 over the next ten years, and to between 200,000 and 300,000 in the late 1980s. However, there was a much stronger emphasis on renovation and modernization.

The achievement of the 1950s and 1960s was not merely in terms of numbers of units. There was a steady improvement in the quality of housing in terms of area per dwelling and per person, and in the number of dwellings with bath, WC and central heating. This improvement continued in the 1970s and, at a slightly lower rate, in the 1980s (Table 4.2). Some commentators of the time were, and

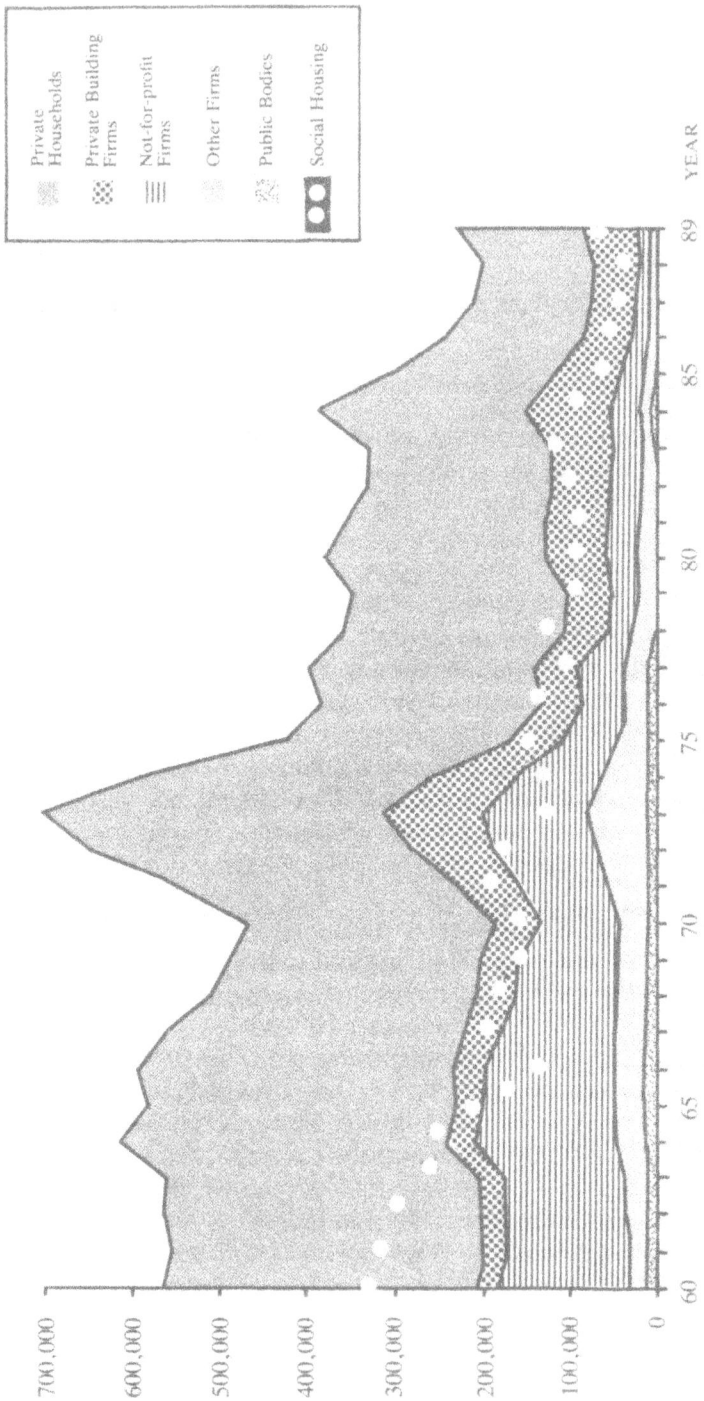

*Figure 4.1* Dwellings completed 1960–89, by sponsors

Legend:
- Private Households
- Private Building Firms
- Not-for-profit Firms
- Other Firms
- Public Bodies
- Social Housing

Y-axis: 700,000 / 600,000 / 500,000 / 400,000 / 300,000 / 200,000 / 100,000 / 0

X-axis: YEAR — 60, 65, 70, 75, 80, 85, 89

some still are (Kennedy 1985) very critical of the conventional low-rise post-war housing, accusing it of being unimaginative, but the tower blocks of the early 1970s proved far more unpopular.

## THE 'PRE-PROGRAMMED' DECLINE OF SOCIAL HOUSING

The extension of the post-war social housing programme to private landlords almost certainly resulted in the production of more social housing than would otherwise have been possible, but it contained a 'time bomb'. After (originally) sixty years, the housing could be decontrolled if the owner repaid the social housing loan. In the 1950s, this deadline seemed a long way off. In any event, the Governments of the time implicitly took the view that the eventual ending of the post-war housing shortage would justify the eventual return of most social housing to the free market. The argument was that the large social housing programme was designed to cope with an acute and abnormal condition of disequilibrium; with the restoration of equilibrium, it would be possible to rely mainly on the market system.

An alternative view, advanced by a few lone voices in the non-profit movement, was that disequilibrium was inherent in the housing market. Unforeseeable shifts in population and demand would always

*Table 4.2* Housing statistics 1968–87

|  | *Units* | *1968* | *1978* | *1987* |
|---|---|---|---|---|
| Housing units | Mill. | 20.30 | NA | 26.60 |
| Households | Mill. | 20.85 | NA | 26.74 |
| Persons | Mill. | 58.98 | NA | 62.40 |
| Private dwellings | Mill. | 19.66 | NA | 26.28 |
| Rented dwellings | Mill. | 12.17 | NA | 15.37 |
|  | % | 63.6 | 62.5 | 60.7 |
| Rent per $m^2$ | DM | 2.28 | 4.32 | 6.37 |
| Area per dwelling | $m^2$ | 61.0 | 66.6 | 69.2 |
| per person | $m^2$ | 22.5 | 29.3 | 33.0 |
| Owner-occupied dwellings | Mill. | 6.98 | NA | 9.95 |
|  | % | 36.4 | 37.5 | 39.3 |
| Area per dwelling | $m^2$ | 89.3 | 103.5 | 112.7 |
| per person | $m^2$ | 25.5 | 33.8 | 38.3 |
| Dwellings with bath, WC, central heating | % | 29.8 | 60.7 | 73.3 |

*Source:* Statistisches Bundesamt, *Gebaeude – und Wohnungszaehlung*, 25 October 1968 and 25 May 1987: *1% Wohnungsstichprobe* 1978

lead to temporary and local 'housing shortages' and, although there was an inherent tendency to equilibrium, the process of reaching it could involve hardship for vulnerable social groups. On this view, a permanent stock of social housing was needed as a 'safe haven'. (There were, of course, also Marxists who advocated a 'centrally planned economy' for all housing. Such views never had much influence outside German Universities.)

### Criticisms of social housing

The 1970s saw a resurgence of neo-liberal views which interpreted a 'socially responsible market economy' more narrowly than previously, and rejected social housing and even the non-profit movement as an imperfection in a market system (Schneider and Kornemann 1977). Even politicians who were not whole-hog free marketeers began to argue that, with the ending of the housing shortage, the scope of public housing policy could be narrowed considerably. They argued that most social housing should be privatized, with 'bricks and mortar' subsidies replaced by the housing allowance. The state should concentrate on removing obstacles to the provision of housing, and on improving the *Wohnumfeld* (the surroundings of individual dwellings). In 1979, a leading Christian Democrat politician and a like-minded economist foreshadowed the policies of the 1981 Federal Government:

> In the last thirty years, the conditions in the housing market have changed fundamentally. In contrast to the post-War period, the provision of housing for the population is very largely assured; the share of owner-occupancy is considerable; and the needs of particularly vulnerable groups are generally provided for. These changes allow the state to withdraw, for the most part, from the housing market and devote itself to its true, and previously neglected functions. In this way, considerable tax savings will be made, and the debt burden on the state and the tax burden on the individual reduced.
>
> (Biedenkopf and Miegel 1979, p. 125, Authors' translation)

The argument that social housing was no longer needed was re-inforced by defects in its administration. The rents for social housing were based on the *original cost* of individual dwellings, plus main-tenance costs; rent pooling was not allowed. Over time, even the low German inflation rate began to cause substantial differences in rents between older and newer dwellings. These differences did not correspond to differences in household income; indeed, the reverse was often the case, with young, relatively poor, households being placed in expensive new housing. New social housing – based on subsidized but fixed-interest finance, and thus suffering from high interest costs in the early years ('front loading') – became relatively expensive. It was sometimes dearer than corresponding private rented housing, where landlords could adopt a more flexible rent policy over time. ('How tenants are cheated by social housing' was the heading of one critique). This problem could have been resolved by allowing associations to pool rents, something which was advocated from the 1960s onwards. However, such a change would have meant higher rents for longer established tenants, and they naturally resisted it. No change was ever made in the system.

Another source of criticism was that people who obtained social housing when their incomes were below the qualifying income could retain the housing even if their income subsequently rose well above the limit (*Fehlbelegung*, 'incorrect occupation'). It was widely felt, however, that, if they were forced to leave, hardship and 'ghettoization' would result. A system of rent supplements for tenants in this situation has now been introduced; this has largely countered the argument that subsidies were going to people who did not need them.

Another problem was that the maximum cost guidelines for the social housing programme often made it difficult to build social housing in the high-cost conurbations, where it was most needed. Social housing also suffered in public estimation as a result of the high-rise 'concrete bunkers' which were built in the late 1960s and early 1970s. Not all the high-rise blocks consisted of social housing, but the association of some well-publicised monolithic schemes with social housing projects of *Neue Heimat* damaged the public perception of both social housing and the non-profit movement.

Finally, it was claimed, with some justification, that the independent non-profit companies often declined to accept the poorest and most needy applicants, preferring people in steady employment. In 1984, the Federal Housing Ministry proposed that all nomination rights should be vested in local authorities. The non-profit companies refused to accept this proposal, wishing to retain the right to refuse to accept 'problem cases'; this contributed to a souring of relations with the Government.

### Assistance to owner-occupation

Together with increased criticism of social housing in the 1970s and 1980s went increased praise for owner-occupation. As early as the late 1950s, the Germans (who, in the cities, had been a nation of tenants since the 'founding years' of the 1860s) were beginning to discover the attractions of the house, as opposed to the flat, and of owner-occupation as opposed to renting. The percentage of owner-occupation was (and at 40 per cent, still is) relatively low, and many housing commentators, not only of a Christian Democrat persuasion, began to argue that owner-occupation ought to be encouraged.

Until 1987, some assistance was available for low-income home-owners through both the social housing programme and housing allowance. The tax arrangements for owner-occupation were, however, supposedly 'tenure neutral', in that there was a tax on 'imputed rent', with an offsetting allowance for mortgage interest. In fact, the tax on imputed rent was based on outdated valuations, so that there was a substantial tax advantage for home-owners. This system was changed in 1987, when the tax on imputed rent and old-style mortgage interest relief were abolished. This 'private good solution' was defended on the ground that one does not pay a tax for the use of a refrigerator or a boat; why should one pay for the use of a house? Rather inconsistently, however, a new tax allowance was introduced. Anyone buying a house, new or old, could deduct up to DM 15,000 p.a. from his/her taxable income for eight years, 'once in a lifetime'. This new system – as with any allowance against taxable income – gives the greatest benefit to households with the highest incomes.

### The decline of social housing

As political support for social housing ebbed, the number of social housing units built dropped from over 300,000 a year in the 1960s to under 50,000 in the late 1980s. Expenditure on social housing was overtaken by tax expenditure for owner-occupation, and expenditure on housing benefit also rose sharply (Table 4.3). But of even greater importance for the longer term was the creeping decontrol of the social housing stock, and the abolition of the non-profit movement.

Social housing loans contained, from the beginning, a provision that, if the owner repaid the loan at the end of the loan period (originally sixty years), the dwelling became decontrolled. Thus the fuse for the time-bomb under privately owned social housing had been burning since the 1950s, but it was shortened by the Social Democrat–Free Democrat government in 1980. The Government reduced the period for which the dwellings had to be retained as social housing; the effect was to reduce the average outstanding period from 30 to 15 years. This legislative change was more

Table 4.3 Public expenditure on housing and related fields 1970–90

| | 1970 | 1975 | 1980 | 1985 | 1986 | 1987 | 1988 | 1989 | 1990$^x_x$ |
|---|---|---|---|---|---|---|---|---|---|
| | | | | | billion DM | | | | |
| Social housing | 4.0 | 4.6 | 6.0 | 5.6 | 6.2 | 6.1 | 6.0 | 5.8 | 6.2 |
| Tax concessions, owner-occupation | 1.5 | 3.3 | 5.5 | 7.2 | 7.1 | 7.3 | 7.1 | 7.6 | 7.3 |
| Tax exemption, non-profit making | 0.2 | 0.2 | 0.2 | 0.3 | 0.3 | 0.3 | 0.3 | 0.3 | 0.2 |
| Modernization and energy saving 1–4 support for housing | 0.2 (5.9) | 0.3 (8.4) | 1.1 (12.8) | 0.8 (13.9) | 0.8 (14.4) | 0.9 (14.6) | 0.9 (14.3) | 0.9 (14.6) | 0.8 (14.5) |
| Premiums for saving with building societies | 2.4 | 3.7 | 2.9 | 1.7 | 1.6 | 1.6 | 1.5 | 1.5 | 0.9 |
| Housing allowance | 0.6 | 1.7 | 1.8 | 2.5 | 3.4 | 3.7 | 3.6 | 3.7 | 3.8 |
| Urban renewal$^x$ | 0.3 | 1.1 | 2.3 | 1.5 | 1.9 | 2.3 | 2.7 | 3.1 | 2.2 |
| TOTAL | 9.2 | 14.9 | 19.8 | 19.6 | 21.3 | 22.2 | 22.1 | 22.9 | 21.4 |
| Total in 1980 DM | 15.1 | 18.7 | 19.8 | 16.2 | 17.6 | 18.3 | 17.5 | 18.2 | 16.6 |

$^x$ Federal and state programmes only
$^x_x$ estimates

Source: Subsidy, social and financial reports of the Federal Government

concerned with short-term finance than high principle. By encouraging the repayment of loans, the Government was able to maintain a larger programme of social housing expenditure than would otherwise have been possible. The stock of social housing has now begun to decline sharply, and will have fallen by about half by the year 2000.

The abolition of the non-profit movement was far less foreshadowed than the decline of social housing, since the movement was, until the 1980s, rarely the subject of any radical critique, and was often held up by both Christian Democrats and Social Democrats as exemplifying their ideals. Indeed, the episode can be regarded as the outcome of a chapter of accidents. As housing completions started to fall in the late 1970s, the private builders began to criticize the building-for-sale of some of the non-profit companies as 'unfair competition', but their complaints made little impact on the Government. Then the financial plight of *Neue Heimat* began to hit the headlines.

### The fall of *Neue Heimat*

*Neue Heimat* was founded by the German Trades Union Congress, and it became an enormous enterprise, eventually owning 300,000 dwellings as well as building extensively for sale. In the early years it made a valuable contribution to housing supply but in the 1970s – as subsequently became clear – it increasingly lost touch with commercial reality, and succumbed to megalomania. It bought large amounts of land in remote sites, in expectation of development which did not occur; it engaged in risky enterprises in Third World countries, for which it was often not paid; it built too many of the kind of housing projects which won architectural prizes in the 1960s; it treated tenants in a high-handed way. As its financial situation gradually deteriorated, the facts were deliberately concealed by senior executives. When the situation eventually became apparent, the Trade Union Congress tried to turn the company around, but it was too late. To make matters worse, some of the senior executives used their position for improper financial gain. In the end, most of the housing was taken over by the *Laender*, using newly created companies.

The ending was happy enough for the tenants, but the long drawn-out saga had a devastating effect on public perception of the whole non-profit movement, even though *Neue Heimat* was untypical in its size, organization, and problems. There were some 1,800 non-profit companies, mostly locally based, with housing stocks averaging around 2,000 – and in total managing ten times as much housing as

*Neue Heimat. Neue Heimat's* problems were the result of its sheer size, combined with poor managerial oversight at a high level, which may have had something to do with the structure of the Trade Union Congress.

Just at this time, the Finance Minister, Herr Stoltenberg, committed himself to reducing the subsidy bill, which had been the subject of international criticism. He tried to cut the subsidies to shipbuilding, but failed; his own constituency relied heavily on shipbuilding! He then turned to the non-profit housing movement, drawing attention to the tax concessions it received – even though they were trifling compared with other tax concessions (Table 4.2). The Government and the Christian Democrat Party was far from united on the issue, but the combination of the *Neue Heimat* 'scandal' and a widespread belief that there was no longer any housing shortage tipped the balance. Moreover, the non-profit movement was not wholly united in opposition. Some of the companies which had branched out into building for sale found the (individual project) 'cost covering' rules restrictive, and felt that they could do better as private companies.

After a, by the standards of the FRG, remarkably short period of discussion, an Act was passed in 1988 which, with effect from December 1989, brought to an end a century of non-profit housing. Within weeks, the Wall fell, and a new housing crisis was unleashed. If the Wall had fallen eighteen months earlier the Act would never have been passed. A year later, the legislation – which had never been discussed with reference to German unification – was imposed on the new *Laender*.

The 'former non-profit companies' have now renamed themselves 'entrepreneurial' companies. Together with the co-operatives, they have linked up with the housing companies in the new *Laender*, and changed the name of their association to *Gesamtverband der Wohnungswirtschaft* (National Association of Housing Enterprises). They hope to continue to play a distinctive role in a united Germany. Can they continue to serve a 'social' purpose, or will they be forced by commercial pressure to operate as profit-maximizing companies? Those companies owned by local authorities, and perhaps the churches, will almost certainly continue to operate social housing, and to cater for 'need' rather than merely 'demand'. However, other companies may find it difficult to diverge from a profit-maximizing mode. If they do so, their shareholders will be tempted to make a capital gain by selling out to more commercial operators. It therefore seems unlikely that all the 1 million dwellings which were owned by the formerly non-profit companies, but were not social housing, will continue to be let at below-market rents. To this extent, the affordability problem will be worsened. It is also unfortunate that the new financial arrangements for the companies give them little incentive to build new housing.

The co-operatives have not been affected by the legislation abolishing the non-profit companies. They, and the successors to the 'co-operatives' of the Democratic Republic, now form a part of the new association. The future of the East German group is unclear, but they could have a successful future, especially if the Federal Government would reduce the debts with which they are burdened. The co-operatives in West Germany will no doubt continue to manage about 1 million dwellings. Some commentators foresee an expansion of co-operatives, offering a democratic, non-bureaucratic form of housing management (Novy 1991). In the past, however, co-operatives have not flourished as well in Germany as in Scandinavia.

## THE EMERGENCE OF A 'NEW HOUSING SHORTAGE'

The change of emphasis in housing policy away from social rented housing and towards support for owner-occupation and individual assistance through the housing allowance was given impetus by the temporary glut of housing in the mid-1970s. This followed the speculative boom which led to the peak output of 670,000 units in 1974. And yet, by the late 1970s, there were increasing complaints that some low-income groups in the population were finding it difficult to obtain accommodation they could afford. The issue was thus one of affordability rather than physical shortage. There was a protracted argument in the early 1980s on the existence of a 'new housing shortage'. On one side were those who argued that the housing market was in balance; that housing conditions were better than ever before and that the housing allowance provided an adequate 'safety net' for low-income households. On the other side were those who argued that some social groups were in fact facing real, and growing, problems. What light do the housing statistics throw on this issue?

### Housing conditions in the 1980s

When we examine the changes during the 1980s, it becomes clear that, with regard to 'facilities' and overcrowding, conditions improved, although more for some groups than others. In all income groups,

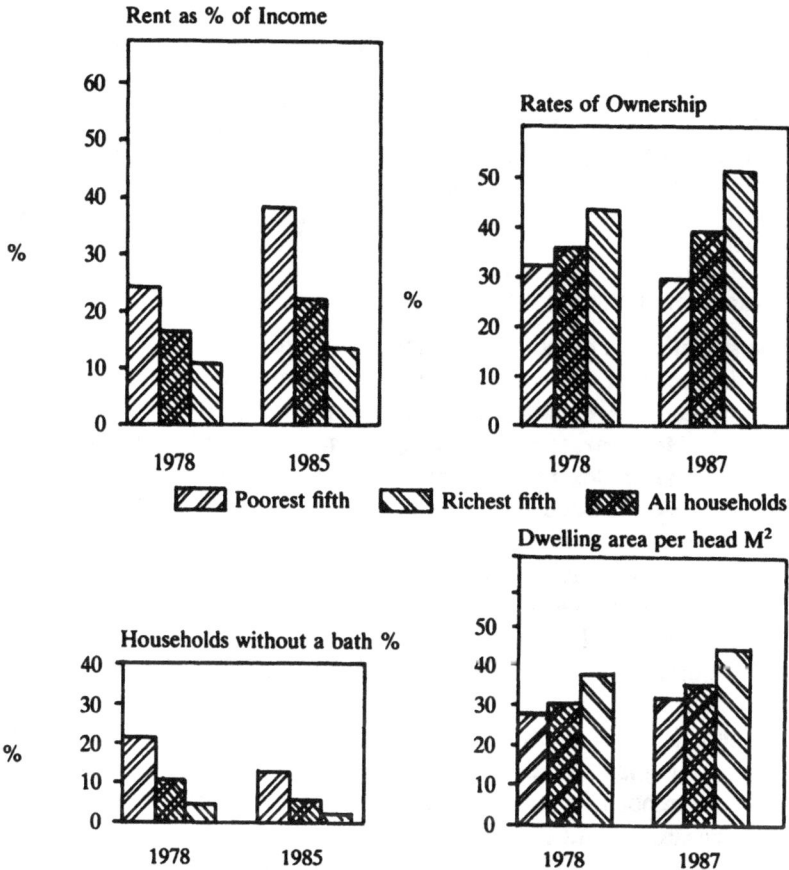

*Figure 4.2*  Housing conditions by income group, 1978 and 1985/7

there were sharp falls between 1978 and 1987 in the percentage of households without a bath, although a substantial difference remained between the richest and the poorest fifth (Figure 4.2).

Space standards also improved. Even for the poorest quintile, the average floor area per head rose from 28m$^2$ to 31m$^2$. In 1978, 17 per cent of households (of two or more persons) were overcrowded, taking the lower standard of less than one room per person; the percentage ranged from 23 per cent in the poorest quintile to 11 per cent in the richest. Because of inadequate statistics, this calculation cannot be made for 1987, but it is possible to compare overcrowding by size of household. Larger households were more frequently overcrowded than smaller households and, over this period, the situation of larger households did not improve; it rather deteriorated. In 1987, 75 per cent of tenant households of five persons or more were overcrowded, and the figure for immigrant households was over 85 per cent. In large owner-occupier households, the figure was only 25 per cent. Similarly, although the average floor space of the poorest fifth rose, there was a slight fall for poor households with five or more persons. In general, we can say that, in terms of facilities and space, the 1980s saw continued improvement, except for large households.

## Owners and tenants

The statistics for owner-occupation and, in particular, rent levels are less encouraging. Surveys indicate that German households prefer an owner-occupied dwelling (preferably a house) to other forms of housing. However, to an increasing extent, this dream can be realized only by households with high incomes; poorer households have been increasingly excluded from owner-occupancy. In the 1970s, a good third of newly-acquired owner-occupied dwellings were acquired by households in the lower half of the income distribution. In the mid-1980s, the figure had fallen to 20 per cent. Between 1978 and 1985 the percentage of owner-occupiers in the bottom quintile actually fell, whereas the percentage in the top quintile rose sharply (Figure 4.2).

There is nevertheless a substantial number of owner-occupiers with modest incomes. In most cases they are elderly people, mostly in one/two person households, who acquired their dwellings a long time ago. In larger households, on the other hand, there is a clear and growing division of society; the richer are owners and the poorer have to be tenants. In 1987, in households of more than five persons, only 10 per cent of the owner-occupiers, but 36 per cent of tenants, fell into the lowest income group. The lower percentage of owner-occupancy among larger, poorer families explains the decline in their (relative) provision with floor-space; as tenants, they have to be content with much smaller dwellings than owner-occupier households.

## The rising rent burden

The trend in rents since 1978 (in the usual German measure of DM per square metre per month) is shown in Figure 4.3, for pre-1939 and

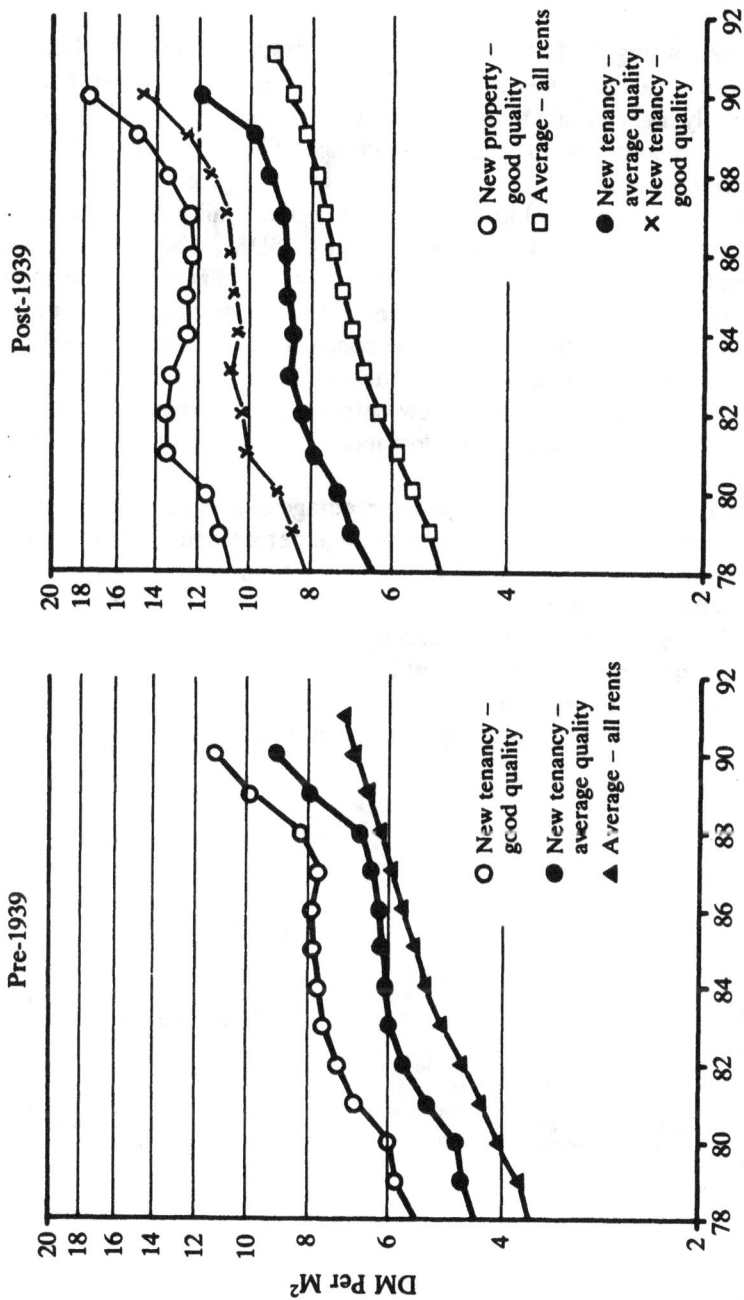

**Pre-1939**

O New tenancy –
  good quality

● New tenancy –
  average quality

▲ Average – all rents

**Post-1939**

O New property –
  good quality

□ Average – all rents

● New tenancy –
  average quality

× New tenancy –
  good quality

DM Per M²

*Figure 4.3* Average rents 1978–91

post-1939 housing. In both categories, the average rent for the whole stock rose at a fairly steady rate. New lettings, however, were at significantly higher rents than the average. Thus in 1990 new lettings of average quality pre-1939 housing averaged 9DM as against 7DM for the whole stock. For post-1939 dwellings, the figures were 12DM and 9DM respectively, while net lettings of good quality 'pre-owned' housing averaged 14DM, and new housing 18DM. The difference between the rents of new lettings and average rents is partly due to legal restrictions on the rate at which rents may be increased, and partly due to landlords making some concessions to sitting tenants. There has been a noticeable acceleration in the rents of new lettings since 1988. In such a situation, new entrants to the housing market – especially young people on modest incomes – often find it difficult to pay for housing.

In the early 1980s, rents rose as a percentage of average household income; between 1978 and 1988 the average percentage rose from 18 to 23 per cent. Of course, a rise in rents (in relation to incomes) may simply reflect an increase in the quantity or quality of housing. The *Institut Wohnen und Umwelt* has attempted to isolate the components in the increase in the average 'rent burden' between 1978 and 1988 (Figure 4.4). It can be seen that the rise from roughly 18 per cent to 21 per cent can be explained by changes in household composition,

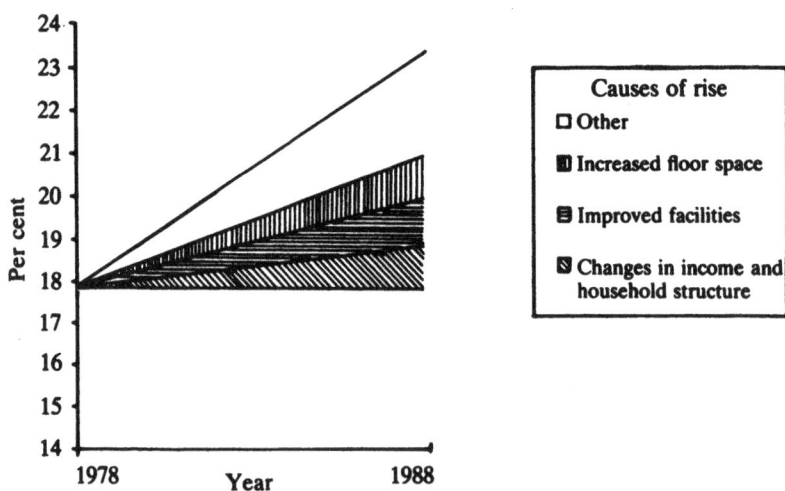

*Figure 4.4*   Causes of rent rises, 1978–88

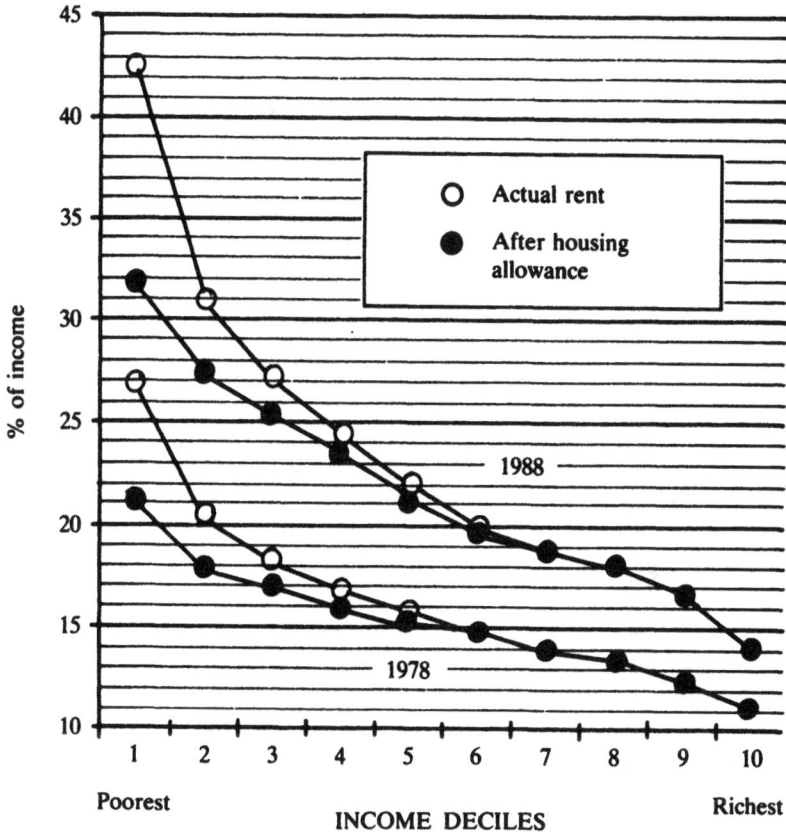

*Figure 4.5*  Rent as a percentage of income by income groups

increased floorspace, and better facilities. However, the remaining 2 per cent – from 21 to 23 per cent – cannot be explained by these factors, and represents a 'price rise'.

A 2 per cent rise in the 'rent burden' may be considered fairly small. However, surveys of household expenditure show that the rise was higher for households on low incomes (Figure 4.5). For households in the poorest tenth, the rise in gross rent (excluding the housing allowance) between 1978 and 1988 was from 27 to 43 per cent. (It must be added that the rise was exaggerated, although not by more than a few percentage points, by a change in statistical methods.) The low-income households particularly affected by rising expenditure on rent include young households and single parents, especially women.

The rising rent burden of low-income households is frequently

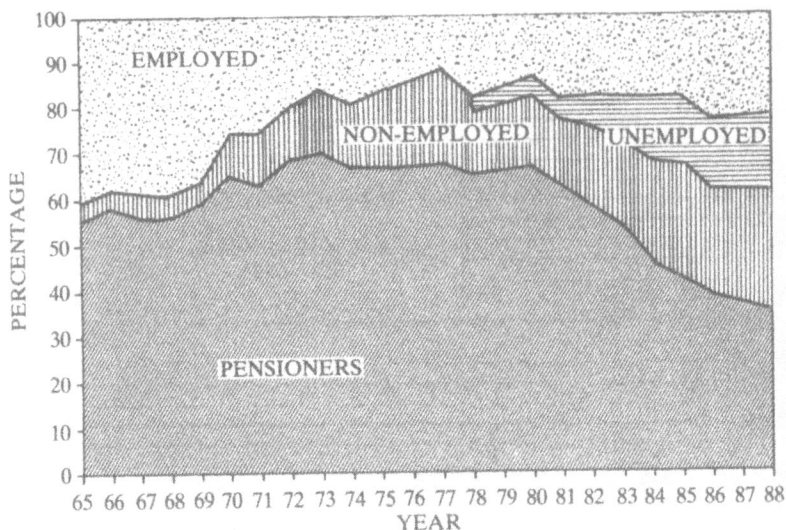

*Figure 4.6*   Recipients of the housing allowance

attributed to the loss of cheap accommodation as a result of de luxe modernization, or conversion to other uses. This explanation is borne out by the fact that rents at the cheaper end of the market have risen more than at the upper end. From 1978 to 1987, rents for the cheapest fifth of rented dwellings rose by 80 per cent, while the rents for the dearest fifth rose by 60 per cent. As poorer families tend to occupy the cheaper dwellings (at least in the private sector), this change in the rent gradient explains why the rent burden of poor households has risen most sharply.

**The housing allowance**

The rise in the rent burden should not, however, be discussed without reference to the effect of the housing allowance (*Wohngeld*). This can have a considerable impact; in 1988, it reduced rent expenditure by 37 per cent for all recipients, and by 60 per cent or more in households with very low incomes. However, more than half of eligible households do not apply for it. Figure 4.5 shows that poorer households benefit most from the housing allowance but still pay the highest percentage of income in rent. In the lowest income decile in 1988, the housing allowance reduced the percentage of household income paid in rent from 43 to 32 per cent; in the next lowest decile, the reduction was from 31 to 27 per cent. In the top four deciles, in which the housing allowance was insignificant, the percentages ranged from 19 to 14 per cent.

The number of recipients of housing benefit has fluctuated between 1.6 and 2 million since 1973, while the expenditure has risen steadily, with a particularly sharp rise from 1985 to 1988. The composition of recipients shows interesting changes (Figure 4.6). Over the past twenty years, employed persons have remained at around 20 per cent of all recipients, while unemployed or non-employed persons have increased sharply to around 40 per cent, and the percentage of pensioners has fallen.

**The numbers of homeless and near-homeless**

Given that a minority of the population did not share in the housing improvements of the 1980s, is it possible to estimate how many people suffer from 'inadequate' housing? This question can be answered in respect of 'homelessness'; of the number of people living in housing without specified facilities – internal WC, bath, central heating – or in physically 'unfit' dwellings or overcrowded conditions; and of persons with an 'affordability problem', based on some estimate of tolerable housing costs in relation to income. Much of the information on these matters dates from before the influx of population into the 'old' Federal Republic which began in 1989. However, a summary of the latest information on the situation in the united Germany is given in Table 4.4.

The category of 'homeless' covers a range of situations. In the old Federal Republic, it is estimated by the Federal Association for Helping the Homeless (BAG) that there may be some 130,000 people, primarily single men, who do not have any fixed abode. A different category are people who would be homeless if they had not applied to local authorities for housing and been placed in temporary 'shelters'. The estimate for 1990 is 300,000, and includes a wider social group, including families with children and foreign families. A further 100,000 people in a similar situation are in cheap bed-and-breakfast accommodation. It is estimated that this total of 400,000 has increased substantially, perhaps doubled, since 1980.

Another group of people are those in homes or institutions. This group, whose members usually have some physical or mental disability, is not necessarily badly housed. However, the trend towards closing 'asylums' creates the need for new 'special needs' housing. Then there are the residents in 'transit camps' (*Uebergangsunterkuenften*), which deal exclusively with German immigrants. The

*Table 4.4* Housing stress in the united Germany 1987–90

| Category | No. persons | Mainly affected |
|---|---|---|
| 1 Old FRG | | |
| Total homeless | (c.800,000) | |
| including: | | |
| fully homeless | 130,000 | Single men |
| in shelters | 300,000 | Families with children, single mothers, single men & women, foreign households |
| in bed-and-breakfast | 100,000 | |
| in institutions | 100,000 | |
| in transit camps | 200,000 | German 'settlers' |
| 1a Potentially homeless, New *Laender* | 200,000 | |
| 2 In unsatisfactory housing old FRG, Total | (800–950,000) | |
| including: | | |
| in hostels | 40–100,000 | Single foreigners |
| dwellings without bath/WC | 500–600,000 | Low-income groups |
| overcrowded (2+ per room) | 350,000 | Large families, foreigners |
| 2a Unsatisfactory housing (former GDR standard) | 300,000 | New *Laender* |
| 3 With excessive rent burden, old FRG (40% + income) | 1.7–1.8m | |

*Source:* Federal Association for Helping the Homeless

estimated total of 200,000 at the end of 1990 has, for obvious reasons, shown a sharp increase.

A second category are people in unsatisfactory housing; the figures are for the old Federal Republic in 1987. It is estimated that there were between 40,000 and 100,000 single foreigners in hostels. In 'normal' housing, between 500,000 and 600,000 people lived in housing without a bath or internal WC and 350,000 in seriously overcrowded conditions, with two or more persons per room.

There are no figures for 'unfit' buildings; this is simply not a problem in the old Federal Republic. Indeed, it can be argued that housing is 'too good', in that excessive modernization has depleted the stock of shabby but cheap housing, which served a useful social purpose. The situation in the new *Laender* is very different. According to the undemanding criteria of the former German Democratic Republic, 300,000 persons lived in unsatisfactory accommodation, but it is now estimated that 700,000 dwellings are uninhabitable, and fit only for demolition.

A final group of persons, in the old Federal Republic, are those with an excessive rent burden. It is estimated that 1.7–1.8 million people are in households where 40 per cent or more of household income goes on housing, although the number falls to around 800,000 when the housing allowance is taken into account. This percentage of income represents an excessive rent burden by any standard but, for people on low incomes, even a much lower percentage payment can involve hardship.

### The current housing situation

To sum up the current situation in the old Federal Republic, we can say that the number of fully homeless is very small, but that the numbers in public makeshift accommodation rose in the 1980s and soared after 1989. It will take several years before the existing immigrants can be accommodated at current standards, and the future course of immigration is unforeseeable. There are small, although significant, numbers of people in housing lacking modern facilities, or living in overcrowded conditions. These problems, however, are (or up to 1989 were) relatively unimportant; in this respect, the old German Federal Republic differs from most other West European countries. The main problem is one of affordability. Households with lower incomes live in older, smaller and less well-equipped dwellings than people with higher incomes, and have not fully shared in the general improvement in housing conditions. The figures on owner-occupancy indicate that households with low incomes cannot keep up. A substantial number of tenants face severe affordability problems. In the extreme case, an inability to finance the cost of housing can cause the loss of one's home, and the subsequent descent into homelessness. The increased rent burden partly reflects a rise in the size and quality of dwellings, but it also reflects a rise in (real) costs, including land and building costs and, especially at the present time, high interest rates. At the same time, the housing situation of the great majority of lower-income households is by no means bad. Nor should the very real problems of a small economically and socially disadvantaged minority make us lose sight of the clear upward trend in the general level of housing provision.

How can we explain the paradox that a section of the population faces severe and growing difficulties in obtaining adequate housing, when there has been such a huge increase in the quantity and quality of housing over the years? There are several strands in any explanation, including changes in the distribution of income, social changes, and poor targeting of housing subsidies.

## THE UNDERLYING CAUSES OF THE 'NEW HOUSING SHORTAGE'

### People in poverty

In the 1980s there was considerable discussion in the Federal Republic of 'the new poverty'; indeed, in the supposedly 'balanced' housing market which prevailed until 1989, the 'housing shortage' could reasonably be considered as an aspect of the poverty problem. Statistics on these matters, however, are sparse. Current statistics are available on unemployment and the number of 'welfare' (*Sozialhilfe*) recipients, but statistics on income distribution are available only up to 1983. Unemployment has remained at around 2 million (6 per cent) since 1981 – a level lower than the OECD average but one which contributes to poverty. Since the opening of the Wall in 1990, a slight fall in unemployment in the 'old' FRG has been more than outweighed by a rise in the 'five new *Laender*'. The number of welfare recipients in the old FRG rose during the 1980s to over 3 million.

One (politically conservative) examination of social changes over the period 1973–83 pointed out that there were two divergent trends. On the one hand, more people joined the middle-income group. On the other hand, the gap between rich and poor increased; for the 3–4 per cent of the population which could be classed as poor, living standards did not increase between 1973 and 1983, as they had done before 1973 (Miegel 1983: 156).

A more recent study of the numbers of people in poverty uses a sample of family budgets from 1963 to 1983 (Hauser and Semrau 1990). This data excludes persons of no fixed address or in institutions, and the very rich. The study defines poverty in relative terms, and uses the criteria of 40 per cent and 50 per cent of average income.

*Table 4.5* The percentage of persons in 'poverty' (below 50% of average income) 1963–83

|  | 1963 % | 1973 % | 1978 % | 1983 % |
|---|---|---|---|---|
| All persons | 14.8 | 5.5 | 6.2 | 7.0 |
| Persons in households headed by: |  |  |  |  |
| employed male | 13.1 | 4.2 | 4.5 | 4.3 |
| employed male under 24 yrs | 5.6 | 2.8 | 3.1 | 12.0 |
| unemployed male 25–49 yrs | 42.8 | 9.2 | 8.6 | 10.0 |
| unemployed female 25–49 yrs | 25.4 | 27.1 | 33.6 | 46.7 |
| Single parents, 2+ children | 45.1 | 12.7 | 23.6 | 43.6 |

*Source:* Hauser and Semrau 1990

On both measures, the percentage of all persons in poverty fell sharply from 1963 to 1973; from 1973 to 1983 it rose, but remained much lower than the 1963 level (Table 4.5).

The aggregate trend since 1973, however, conceals marked differences between household types. For young people, unemployed people, single parents and couples with large families, the percentages are higher, and have risen much more sharply since 1973. On the other hand, there has been no significant rise among people aged over sixty-five, or even over fifty. The groups experiencing greater poverty since 1973 have also had difficulties of access to housing – young people because they are *ipso facto* newcomers to the housing market, and parents of large families because, in Germany today, they are a deviant group.

Hauser and Semrau conclude that, up to 1983, there was no evidence of the growth of an economically depressed 'underclass', but that, in the previous decade, the position of certain groups, notably young people and parents of large families, had worsened, and necessitated a change of emphasis in anti-poverty policy. How far the situation of these groups has changed since 1983 cannot be demonstrated statistically, but impressionistic evidence suggests that it has not improved.

### Social changes

Other factors behind the continuing 'housing shortage' are of a social nature; one is a change in the structure of households. The average household size has halved over the past hundred years – from 4.5 to 2.25; in 1950 it was 3. That means a doubling of the number of dwellings for a given population. Over this hundred years, the percentage of single-person households rose from 7 per cent to 35 per cent (in 1950 it was nearly 20 per cent), while in the large cities it is often over 50 per cent. In this respect, therefore, the pressure is strongest in the large cities. The fall in household size is continuing, since young people leave their parents earlier, and marry (or form relationships) earlier than in the past, leaving their parents in a 'family dwelling'. The increased divorce rate often has the same effect. Moreover, as a result of affluence and less gregarious habits, unused rooms which in the past would have been let, are today often left unused.

A further factor tending to increase demand for housing has been the continuing high (and unequal) growth of real income, which was not foreseen at the beginning of the 1980s, and which has contributed to the insufficient level of housebuilding. The growth of real income has led to an increase in the size of new dwellings and to some amalgamations of small old dwellings, as well as the growth in the number of second homes.

### The poor targeting of housing expenditure

Finally, housing subsidies are poorly targeted, if the aim is to help people in need. Table 4.3 shows the development of the various categories of public assistance for housing. Out of the total of some 20 billion DM, a third goes to tax relief for owner-occupation and a quarter to 'social housing'. About one-fifth goes into 'urban renewal', although most of this is spent on town planning measures (roads, parking etc.) and only 4 per cent on the modernization of dwellings themselves. A further sixth is spent on the housing allowance. When one examines the changes in expenditure over time, it is clear that the three categories which have grown are: tax allowances for owner-occupancy, the housing allowance, and town planning (urban renewal) expenditure. On the other hand, the share of social housing and of premiums for saving through building societies (*Bausparfoerderung*) has fallen.

The different types of expenditure have different impacts on the housing situation of the less well-off. Social housing expenditure creates new housing, and ensures that it is available for lower (although not always lowest) income groups. The housing allowance does not in itself create any new housing, but it is well targeted from a social point of view. As Figure 4.7 shows, 70 per cent of the amount spent in this way benefited the bottom fifth of the income

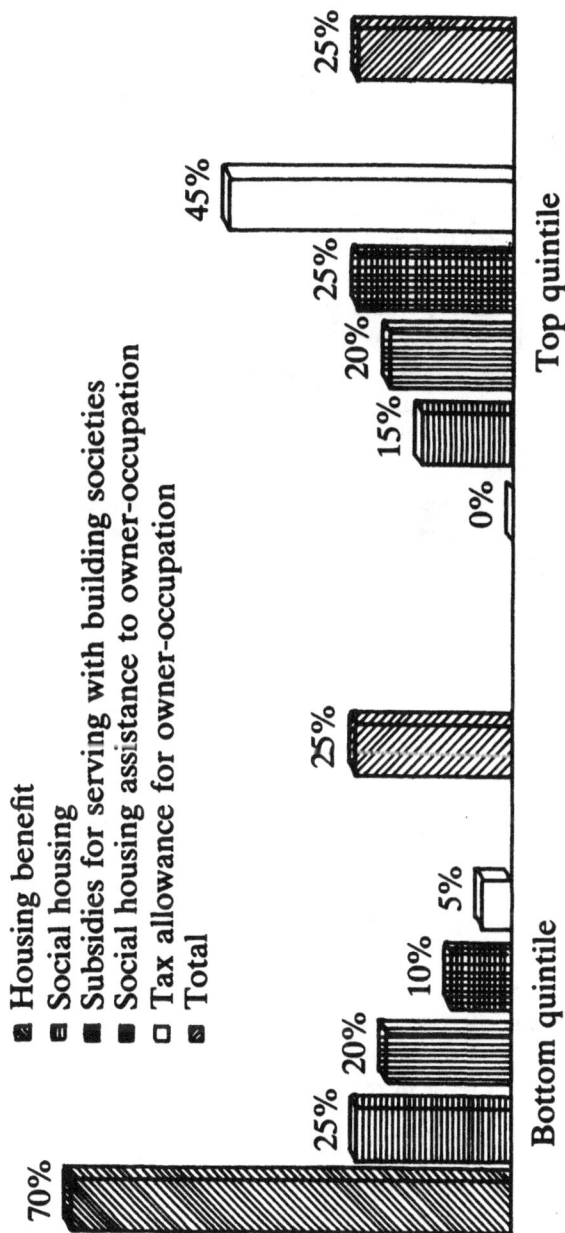

**Legend:**
- Housing benefit
- Social housing
- Subsidies for serving with building societies
- Social housing assistance to owner-occupation
- Tax allowance for owner-occupation
- Total

**Bottom quintile**

70% | 25%
20%
10%
5%
25%

**Top quintile**

0%
15%
20%
25%
45%
25%

*Figure 4.7* The distribution of housing subsidies between households of the top and bottom income quintiles

distribution. Social housing, as well as the premium for saving with building societies, benefited this group only slightly more than the average (25 and 20 per cent respectively).

On the other hand, the top quintile of the income distribution benefited enormously from the tax concessions to owner-occupancy; 45 per cent of the total went to this group. Thus the top and bottom quintiles both received 25 per cent of the total 'cake' i.e. somewhat more than their proportionate share, while the middle three quintiles received somewhat less than their proportionate share. If we allow for secondary effects, e.g. filtering, we can say that the distribution of housing subsidies was broadly 'neutral' in its distribution among the various income groups. In other words, the total expenditure on housing did not achieve any transfer from rich to poor, and to this extent served no social purpose at all!

It should be added that these are average and not marginal figures. Under different arrangements, marginal expenditure on social rented housing or even assistance for owner-occupation could, in theory, be targeted more effectively on lower income groups. However, the housing subsidy system, as it has actually operated, has been very inefficient in tackling the serious problems of affordability and availability which – in spite of a generally excellent housing situation – prevail in the bottom third of the income distribution, as well as in special groups such as large and single-parent families.

## REBUILDING A NATIONAL HOUSING POLICY

The achievements of the post-war 'housing miracle', the continued improvement in housing conditions, and the forecast of a falling population led to a false complacency in the FRG in the 1980s. It was too readily assumed that the 'housing problem' was over for good, and that all that was needed was to maintain the housing allowance and fiscal encouragement to owner-occupation. The growing affordability problems of a section of the population were overlooked, and the total output of housing was allowed to fall to a level which, at least in retrospect, was too low. Legislative changes were made which will now inevitably lead to a sharp decline in the stock of social housing. The non-profit movement, instead of being adapted for a continuing role in meeting changing, but continuing, housing problems, was abolished. Not all the changes were ill-judged; a shift in emphasis away from the 'numbers game' of the 1950s and 1960s was necessary. Nevertheless, the balance struck by housing policy in the 1980s was being widely questioned, and the

Federal Government had already begun to change course, even before reunification.

Reunification, and the collapse of communism in Eastern Europe, has profoundly altered the housing situation in the Federal Republic. The extra 7 million dwellings in the new *Laender* raise problems very different in scale and character from those in the 'old' Federal Republic. Many of them are in a catastrophic state, and require huge investment. Rents are far too low, but incomes and savings are also low, and too rapid a rise in rents could create severe affordability problems. There is an urgent need for public action; this is not a problem which can simply be left to the market.

At the same time, the housing market in the old FRG has been put under severe pressure by immigration of 'East Germans' (*Ueber-siedler*), ethnic Germans from Eastern Europe (*Aussiedler*), and asylum-seekers from all over the world. Since the fall of the Wall, the population of the old FRG has increased by some 2.5 million. Assuming unchanged occupancy rates, this increased population would require a doubling of the mid-1990s level of housebuilding for five years. In late 1991, the level of housing starts was some 50 per cent higher than in 1975, and limited by capacity bottlenecks. Average rents were around 10 per cent higher than in 1989, but the rents of new tenancies had risen more. Thus the 'new housing shortage' which – we would argue – already existed in 1989, has been aggravated by the addition of a physical shortage comparable in nature (although not in severity) to that which prevailed after the war. It is, moreover, impossible to predict the numbers of, in particular, *Aussiedler* over the next few years.

The actual and potential immigration into the Federal Republic means that many of the assumptions which underlay housing policy in the 1980s no longer apply. It was assumed that the population of the 'old' Federal Republic would fall by some 2.4 million by the year 2000. On this basis, it was calculated that a construction level of around 250,000 a year would be sufficient to cope with demolition and the continued growth in the number of households, and that large-scale settlements would not be needed. Indeed, in 1985, the legal basis for planning new settlements was repealed (Hallett and Williams 1988, p. 43). These assumptions will now have to be abandoned; land and housing policy will have to return to the post-war ethos of facilitating increased construction of new housing in a variety of ways. At the same time, the recurring idea that 'the housing shortage would be ended if we had 'x more houses' is probably an illusion. An improvement in the housing situation resulting from increased supply could conceivably attract more immigrants, thus worsening the situation.

Nevertheless, changes are needed which will facilitate the supply of land and capital for new housing. To provide more land does not mean abandoning land-use planning, but it does mean combined public and private action on suitable sites. There is now, for example, a considerable amount of redundant industrial and military land which could be used for housing, but only if the financial problems associated with pollution and land clearance (*Altlasten*) are resolved.

Investment in housing by individuals and companies – and of a kind which does not exclude low-income households – could be encouraged by various changes in fiscal policy which have been advocated for many years. Recent tax policy has tended to favour the short-term speculator, who renovates property and resells it, as against the long-term investor. Insurance companies could be encouraged to put more of their funds into rented housing; it provides no quick profits but is an ideal backing for long-term insurance policies. Large firms could be encouraged to provide housing for their employees. The tax (and subsidy) incentives to owner-occupation could be redirected towards moderate-income households, who are on the threshold of owner-occupation, and away from higher-income households who would be owner-occupiers anyway. Self-building and home improvements could be encouraged.

A change of emphasis in housing policy will require expenditure and legislation at the Federal level, but also action at the regional and local level; the constitution of the Federal Republic is fortunately well suited to local initiative. So far, the Government's response to the new housing challenge has been inadequate. It has allocated DM2.2 billion for a new programme of social housing (some of which may be earmarked for the new *Laender*). The money will be used for a new 'Third Way' of housing subsidies (as distinct from the traditional First and Second Ways). This involves giving builders a subsidy for seven years, in return for corresponding rent reductions. After seven years, the dwelling will be decontrolled. The system aims to stimulate the construction of housing to let, and may well do so. But this social housing will self-destruct even more rapidly than the traditional type of privately-owned social housing.

The new situation raises both old and new issues in housing policy. A new issue is that, even with regard to housing, the Federal Republic has a direct interest in encouraging economic development in Eastern Europe. An old issue is that, since a mere expansion of the housing stock, together with the housing allowance, is unlikely to solve affordability problems, more emphasis should be placed on maintaining a stock of moderate-price housing, reserved indefinitely for households with low incomes. In other words, Germany needs non-profit housing! Any such suggestion is clearly unrealistic at the present time but perhaps non-profit housing will be rediscovered in the twenty-first century.

## REFERENCES

Biedenkopf, H. and Miegel, M. (1979) *Wohnungsbau am Wendepunkt*, Stuttgart: Bann Aktuel.
Hallett, G. (1977) *Housing and Land Policies in West Germany and Britain; A Record of Success and Failure*, London: Macmillan.
Hallett, G. and Williams, R. 'West Germany' in Hallett, G. (ed.) (1988) *Land and Housing Policies in Europe and the USA; A Comparative Analysis*, London: Routledge.
Hauser, R. and Semrau, P. (1990) 'Polarisierungstendenzen in der Einkommensverteilung', ISI, *Informationsdienst Soziale Indikatoren*, 3 January, Zuma.
Kennedy, D. (1984) 'West Germany' in Wynn, M. (ed.) *Housing in Europe*, London: Croom Helm.
Miegel, M. (1983) *Die Verkannte Revolution*, Stuttgart: Bonn Aktuel.
Novy, N. (1991) 'Introduction' in Norton, A. and Novy, N. (eds) *Low Income Housing in Germany and Britain*, London: Anglo-German Foundation.
Schneider, H.K. and Kornemann, R. (1977) *Soziale Wohnungsmarktwirtschaft*, Bonn: Eicholz.

# 5 Housing problems in the former German Democratic Republic and the 'New German states'

*Otto Dienemann*

(Written in late 1990 and early 1991)

## THE INITIAL SITUATION AND THE POLITICAL BACKGROUND

When the German Democratic Republic was founded in 1949, the economic situation was desperate. The productive capacity of the country had been devastated during the war, and the division of Germany cut the country off from most of its previous coal, steel and cement supplies. In addition, 680 large enterprises and 7,500 km of railway were dismantled and shipped as reparations to the Soviet Union. Some 10 per cent of housing had been destroyed.

The economic (including the housing) policy of the new state was based on strictly centralized economic planning, with five-year plans linked to yearly plans. The guidelines for each five-year plan were laid down at the congress of the United Socialist Party of Germany (SED, i.e. the Communist Party and some puppet parties). The plans initially gave priority to the building-up of basic industries. Only the bare minimum was done to rebuild the destroyed cities and expand the supply of housing, and the housing shortage soon led to considerable dissatisfaction among the population.

In 1949 most dwellings were privately owned. All dwellings were at once brought under state control, but construction for owner-occupation was readily granted in the early 1950s, and accounted for nearly half of new construction. At the end of the 1950s, however, a new pattern was established. More emphasis was laid on housing construction, and industrialized 'system building' was introduced, using storey-height wall panels for the construction of blocks of flats. State housing for renting was given the lion's share of new construction, followed by cooperative housing, with private construction being confined to less than 10 per cent. (The 'cooperatives' usually catered for employees of individual state enterprises, and differed only slightly from state housing).

The allocation of housing was comprehensively controlled. House-holds seeking accommodation had to apply to the local housing department, and were allocated to both state housing and private tenanted housing. There was no market for either buying or renting. Waiting lists were sometimes long, but allocation was generally on objective indicators of 'housing need'.

The GDR was a racially and culturally fairly homogeneous state. As a result of strict immigration control, there were no significant immigrant communities (apart from some fairly segregated groups of workers from Mozambique and Vietnam). Differences in income were relatively low, and there was full employment, although this concealed extensive 'disguised unemployment', in the sense that many employees did not produce anything of value.

In the late 1950s, in conjunction with the changes in the organiza-tion of the construction industry, the SED decided to give greater priority to housing in the five-year plans. The construction of new housing rose sharply, from an average of 34,000 p.a. between 1950 and 1956 to around 85,000 p.a. in the 1960s. However, at the tenth conference of the SED in 1973, the decision was taken to embark on an even more ambitious long-term housing programme. In the period from 1976 to 1990, 2.8 to 3 million dwellings were to be constructed or modernized, at a cost of 200 billion Marks. The existing plan for 1971–5 was expanded, with the objective that: 'With the construction of 3.3–3.5 million dwellings in the period 1971 to 1990 every household should have a dwelling which is in good physical condition and is of an appropriate size'.

It soon became apparent, however, that the 'made in GDR' economic system was not able to match the productivity and dynamism of the leading Western industrial countries. The SED leadership began to seek socially-orientated arguments for the superiority of its form of socialism. The aims of the Housing Programme were reformulated in terms such as:

> The historical social and geographical differences in living condi-tions are to be eliminated. In this way a basic ideal of socialist development, the gradual attainment of equality between different classes and social groups, will be realised.

> Stable and low rents, independent of income, allow every family and every citizen to obtain and retain a suitable dwelling, and underpin the security of their home for their entire life.

> By carrying out a unified housing programme covering new building, reconstruction and modernisation, the housing conditions

of the maximum number of citizens will be improved, and the renewal of towns and villages will be achieved.

The published aim of government policy was thus a socially just housing system, which would not in any way seek to iron out differing needs, but would provide a variety of types of housing, for large families as well as elderly or single people, for shift workers, scientists and artists, and for cooperative groups, while avoiding the 'ghettoization' of capitalist countries. The promise was of good housing for everyone, not luxury housing for the few.

## WHAT WAS ACHIEVED?

### The total housing situation

The Housing Programme can claim some successes. The annual production of dwellings was raised from 65,000 in 1971 to over 120,000 per annum in the 1980s. Since 1971 the stock of dwellings has been increased by nearly a million. The number of dwellings per thousand population was 259 in 1971, it is now over 425.

There are at present (1991) around 16.2 million inhabitants in 'East Germany' and about 6.5 million households. This represents a fall from a population of 18.4 million in 1950, as a result of a low birthrate and emigration. For every 100 households there are 108 dwellings. For every inhabitant there is an average of 1.2 rooms (living and bedrooms). The average floorspace per person is 27.6 m$^2$ (including all rooms) which is good even by international standards. Moreover, the provision of baths, internal WCs and modern heating has been markedly improved. One of the most important aspects of this improvement has been that it has been accompanied by a reduction in differences between households in housing standards; in nearly all socio-demographic groups, there has been a convergence of housing standards.

On the other hand, there are considerable disparities and inefficiencies in the way the available living space is used. A great deal of

*Table 5.1* Dwelling stock 1950–90

|                              | 1950 | 1961 | 1971 | 1981 | 1990 |
|------------------------------|------|------|------|------|------|
| Dwelling stock(m)            | 4.67 | 5.51 | 6.06 | 6.56 | 7.00 |
| Dwellings per 1000 population | 276  | 320  | 355  | 401  | 426  |

Table 5.2 Housing conditions 1961–90

|  | 1961 | 1971 | 1981 | 1990 |
|---|---|---|---|---|
| Floor space per inhabitant (m²) | 16.8 | 20.6 | 24.9 | 27.6 |
| Dwellings with: | % | % | % | % |
| Bath/shower | NA | 22 | 39 | 82 |
| Inside WC | NA | 33 | 39 | 76 |
| Hot water | NA | NA | 26 | 82 |
| Modern heating | NA | NA | 11 | 47 |

space is under-used. This applies particularly to elderly people, who enjoy low rents and security of tenure, and therefore remain in large dwellings when the children have left. There are also a considerable number of individuals who, perhaps as a result of moving, have been allocated two dwellings, and who retain them. This misallocation of housing space is the outcome of extremely low rents.

**Low rents fixed by the state**

In the GDR, rents were fixed by the authorities, irrespective of whether the dwelling was publicly, privately or cooperatively owned. The rent level varied according to the size and equipment of the dwelling, but was always very low. In Berlin, rents per square metre ranged from 0.8 to 1.25 Marks, although in some central areas they rose to 1.87 Marks. There was a surcharge of 0.4 Marks for centrally or district heated dwellings. In cooperative dwellings the rents, or charges, were even lower, since the dwelling was partially financed by the occupiers.

In general, rents made up less than 3 per cent of household income. Rents for agricultural workers were exceptionally low, at 0.9 per cent. On the other hand, some households with low incomes paid significantly more; households with incomes of less than 1,200 Marks a month paid 4.5 per cent; pensioners with an income of less than 400 Marks a month paid 8.2 per cent; and single men or women with many children and a very low income paid 10 per cent. The low rents were made possible by massive housing subsidies of around 20 billion DM annually. Two-thirds of the cost of housing maintenance was paid for by the state (i.e. rents contributed nothing towards capital costs, and covered only one third of the often insufficient maintenance expenditure).

**Equalization of housing conditions**

The 'Housing Programme' was unable to attain its objective of similar housing conditions in all parts of the country. The centrally planned economy was accompanied by an unjustified preferential treatment of Berlin and certain regional capitals, so that geographical differences were not reduced. As a result of this policy, Berlin had by far the best housing conditions.

Since 1971, the differences between the best and the worst districts, in the number of dwellings per 1,000 inhabitants and the average floor area per head, have increased slightly. On the other hand, the differences in the provision of a bath/shower have fallen markedly. The average figure is 82 per cent and, except for Berlin (97 per cent), there is little variation.

*Table 5.3* The best and worst districts 1971, 1990

|  | Year | Maximum | Minimum | Difference |
|---|---|---|---|---|
| Dwellings | 1971 | 427 (Berlin) | 313 (Rostock) | 114 |
| per 1000 | 1990 | 484 (Berlin) | 381 (Neubrandenburg) | 131 |
| Floor area | 1971 | 24.1 (Berlin) | 17.5 (Rostock) | 6.6 |
| per head, $M^2$ | 1990 | 30.8 (Berlin) | 24 (Rostock) | 6.8 |
| Bath/ | 1971 | 59 (Berlin) | 29 (Chemnitz) | 30 |
| shower, % | 1990 | 94 (Berlin) | 78 (Dresden) | 16 |
| Inside | 1971 | 80 (Berlin) | 22 (Chemnitz) | 58 |
| WC % | 1990 | 95 (Berlin) | 61 (Dresden) | 34 |

In terms of physical condition, however, housing in the small – and middle-sized towns is markedly worse than in the larger cities. The neglect of the small towns is reflected in the dilapidated state of much housing, and in the poor standard of water and sewage facilities. The poor condition of housing and the urban environment reaches its peak in the industrial connurbations of Halle/Leipzig, Dresden/ Upper Elbe Valley/ and Chemnitz (Karl-Marx-Stadt!)/Zwickau. Especially in the smaller communities in these connurbations, living conditions are markedly under the average for the country. In the districts of Dresden, Chemnitz, Gera and Suhl, the percentage of dwellings without an internal toilet ranges from 63 to 69 per cent, well below the average of 72 per cent. There are also significant spatial differences in the provision of other infrastructure. The number of hospital beds per 10,000 inhabitants ranges from under 90 to over 110; the number of medical practices from under 18 to over 20; the number of telephones per 100 dwellings from under 11 to over 15.

*Table 5.4* Equipment with public facilities 1990

| Inhabitants | Water % | Drains % | Sewage Treatment % |
|---|---|---|---|
| <10,000 | 88 | 56 | 36 |
| 10,000–20,000 | 92 | 70 | 36 |
| 20,000–50,000 | 96 | 80 | 62 |
| 50,000–100,000 | 97 | 90 | 76 |
| >100,000 | 100 | 95 | 97 |

**The housing conditions of elderly and handicapped citizens**

The level of comfort in old people's housing is below average, and there are substantial regional differences. The average waiting time for purpose-built old people's 'sheltered' housing is more than five years, and only 15 per cent of the population has been covered. Similarly inadequate is the provision of old people's homes and nursing homes; for every 1,000 pensioners there are on average only sixty-six places.

Most old people, however, would prefer to remain in their own homes. The type of housing modification which makes this possible has, however, been provided for very few. As in the past, the old and socially handicapped live in decaying older housing areas. The situation is particularly depressing for mentally handicapped people, who cannot live without assistance and supervision. Only 12.7 per cent of the persons who, because of mental handicap, quality for special work places are accommodated in sheltered homes or communal dwellings. (There was a tendency among the political leadership to believe that mental handicap, or at least mental illness, would not be as serious a problem under socialism as under capitalism.)

**Dwelling size and the provision of social facilities in new housing areas**

The average floor area for new multi-storey flats has been set at $58m^2$. For 'houses' (which constituted only 10 per cent of total production) the average floor area has been $100m^2$. The prices of new housing were also laid down by the State. They were around 60,000 Marks at the beginning of the Housing Programme, but later rose to 80,000 – 90,000 Marks. As a result of this limit, and the fact that the political leadership equated success with the number of completed dwellings, production was not matched to demand. Too many one- and two-room (i.e. one bedroom) dwellings were built and the dwellings were too small. In the suburban areas where 70–80 per cent of new

*Table 5.5* Types of new housing

| Dwelling type | % of production | Average size (m²) |
|---|---|---|
| 1-room | 18 | 27–39 |
| 2-room | 23 | 39–52 |
| 3-room | 43 | 52–68 |
| 4-room | 15 | 68–84 |
| 5-room | 1 | 84–96 |

*Table 5.6* Social facilities 1971, 1990

| | 1971 | 1990 |
|---|---|---|
| Classrooms per 1000 pupils | 39 | 59 |
| Creche places per 1000 children | 201 | 849 |
| Nursery places per 1000 children | 645 | 1076 |
| Dental surgeries per 1000 inhabitants | 1.0 | 1.9 |

building took place, the average inhabitant had only 20 to 25m² of floorspace. Table 5.5 illustrates the predominance of small dwellings in the Housing Programme. The demand from families for 4–5 room dwellings, tailored to their individual needs, was unsatisfied.

On a more positive note, it must be added that, in the new peripheral housing estates, social facilities and shops have been built at the same time as housing. Thus in some districts the level of social facilities has markedly improved. However, the provision of the new housing areas with indoor swimming pools, sport halls, restaurants and other leisure facilities is generally quite inadequate.

The smallness of the apartments and the lack of facilities in most of the new housing areas encouraged many people to build 'weekend houses' or 'dachas' in areas outside the cities. Some 1.7 million dwellings, ranging from shacks to substantial bungalows, were built in this 'informal sector', which was tolerated by the authorities.

**The decay of the cities and settlements**

The myopic concentration on the new housing areas and the neglect for decades of the older 4–5 storey blocks of flats which predominate in the towns has contributed to a degree of dilapidation in the central urban areas which is apparent to any observer. Particularly affected are small- and middle-sized towns with 10,000 to 50,000 inhabitants, which house 65 per cent of the population. Table 5.7 illustrates the condition of the pre-1945 multi-dwelling housing (with an average age of 86 years); over half requires complete renovation or is

*Table 5.7* The condition of pre-1945 housing

|  | Condition categories (Share in %) | | | |
|---|---|---|---|---|
|  | *1* | *2* | *3* | *4* |
| 1 & 2-family houses | 30 | 50 | 17 | 3 |
| Multi-dwelling housing | 9 | 40 | 40 | 11 |

Category 1: good condition
Category 2: no structural damage, only maintenance needed
Category 3: extensive damage, complete renovation needed
Category 4: uninhabitable

uninhabitable. The one-or-two-family houses, which also have an average age of 83 years, are in much better condition.

In many towns, there is a crass contradiction between their social and cultural importance and their physical decay and obsolescence. A study of the central areas of 70 of the 227 'county towns' (*Kreisstaedte*) found that 30–35 per cent of residential buildings were in housing categories 3 and 4, i.e. either uninhabitable or needing complete renovation; 70–80 per cent of the dwellings did not possess a bath or shower. The decay of these historic towns is a tragic loss to European cultural inheritance. The central areas of more than 200 towns can be classified as of major importance in the history of European architecture. In the period since 1971, 10 to 15 per cent of historically important buildings have been lost; many other buildings and districts are so dilapidated that there is little prospect that they can be restored.

The excessive development in peripheral areas has led to an extension of the urban area by about 30 per cent, and to depopulation and decay in the inner areas. The population density in the inner areas of many towns has fallen by 20–40 per cent; the overall density has fallen by 10–15 per cent. The social decay of central areas has been accelerated by a drastic deterioration in the building condition of shops, restaurants, libraries etc., many of which have closed. The cause has been that the level of maintenance expenditure has been held for years at a level only 30–50 per cent of that needed to keep the buildings in good condition.

### There is still a 'housing shortage'

In 1990, there was still a housing waiting-list of 800,000, of which about one-third are socially urgent cases. Out of around 7 million dwellings, 860,000 were without a bath or shower, 1.3 million were without an inside toilet, and over 3.2 million were without modern heating. Over 65 per cent of the main roads were in structural categories 2 and 4 (poor or very poor) and only 14.5 per cent of communities had a sewage treatment plant.

Moreover, contrary to the assertions of the previous Government, the housing system was not in all respects socially just. A small group of leading Party officials was given highly privileged treatment in the provision of housing and other services. On the other hand, the private owners of the older blocks of flats, and of one- & two-family houses were severely discriminated against, being subject to completely uneconomic rent controls, and denied any public assistance in the maintenance of their property.

The overriding priority given to the numbers of new housing units allowed little scope for rational and humane town planning. There was too much building of very high blocks of peripheral developments, and too little attention to the maintenance of historically important older housing. These negative aspects of the policies adopted weigh heavily in the balance against the positive aspects; that the right to housing was legally guaranteed, and that, according to the prevailing definitions, there was no homelessness.

## THE CAUSES OF THE SHORTCOMINGS

### The inefficiency of the economic system

In the GDR there was at first a widely shared belief in the superiority of a socialist economic system which had eliminated the market and private ownership of the means of production. However, by the 1970s, intellectuals were becoming increasingly aware that the Western industrial countries were outstripping the socialist countries. The gap in labour productivity between the GDR and the leading Western countries grew steadily larger, and reached 40 to 50 per cent. The leaders of the Party and the Government were kept informed of this development by the academic institutes, but the official reaction was, 'Anything that runs counter to socialist ideology cannot be true'. As a result, the published national income statistics of the 1970s and 1980s showed rises in real income which were not in fact achieved. In recent years, the published figures showed growth rates of 3.5–4.0 per cent, although the actual figures were 1–2 per cent. It can now be said openly that the centrally controlled economy of the GDR lacked the dynamism necessary for the successful realization of the highly ambitious 1973–90 'Housing Programme'.

**The tragic consequences of the widespread liquidation of private property**

Soon after 1949 came the first wave of the expropriation of private firms and their conversion into 'public property'. There were eleven such waves in total, before the process ceased in 1972. In manufacturing industry the number of firms fell from 11,560 in 1970 to 3,423 in 1987. Similar developments took place in the building industry, in which the number of independent enterprises was reduced from 970 in 1960 to 190 in 1990/91. The new structure comprised, by output:

72 per cent publicly owned enterprises
15 per cent cooperative craft enterprises
5 per cent private building material enterprises
8 per cent private craft enterprises

The house-building industry was thus dominated by a small number of large firms (*Kombinate*) which covered all aspects of production. Small firms were squeezed out; thus the contribution of private craft enterprises (plumbers, plasterers etc.,) was reduced from 38 per cent of total output in 1950 to 8 per cent in 1989. In this respect, the development in the GDR was quite different from that in the Federal Republic, where the majority of workers were employed in small or middling firms, with close personal and economic contacts to the communes and other employers. As a result of this concentration, many towns and communities lost their local building capacity completely, or retained only specialized sub-contractors. This contributed to the decay of many towns.

The destruction of many small firms and the creation of large combines with thousands of employees led to a dictatorial style of management and 'architectural monoculture'. The concentration on a restricted style of concrete construction, which was insufficiently

*Table 5.8* The structure of the building industry 1989

| *Employees per firm*<br>*(Share in %)* | | |
| --- | --- | --- |
| | *GDR* | *FRG* |
| under 49 | 10.3 | 53.5 |
| 50–100 | 13.4 | 15.3 |
| 100–199 | 4.7 | 13.4 |
| 200–499 | 11.3 | 10.4 |
| +500 | 60.3 | 7.4 |
| Total | 100.0 | 100.0 |

*Table 5.9* Housing ownership 1950–90

|                    | 1950 % | 1960 % | 1971 % | 1990 % |
|--------------------|--------|--------|--------|--------|
| Private ownership  | 91     | 83     | 62     | 41     |
| Cooperatives       | 1      | 2      | 10     | 17     |
| Public housing     | 8      | 15     | 28     | 42     |
| Total              | 100    | 100    | 100    | 100    |

flexible to meet the range of housing needs, was encouraged by the change in the structure of the building industry.

Further unfortunate consequences followed from the policy of reducing the private ownership of housing and land, and placing intolerable burdens on the private owners who remained. Owners of rented property were unable to maintain their property, because of the extremely low rents which were set, while owner-occupiers also found it very difficult to do so, because building materials and labour (monopolized by the building combines) could not be obtained.

**The emasculation of the local authorities**

With the founding of the GDR and the accession to power of the Socialist Unity Party (SED), the political system was profoundly altered. Following a decision of the People's Assembly in 1952, the *Laender* Governments (Mecklenburg-Vorpommern, Brandenburg, Sachsen-Anhalt, Sachsen and Thueringen) which, together with the communes, had been responsible for administration before 1933, and had remained important even under National Socialist rule, were abolished. In their place a system of 14 governmental districts and 227 'counties' (*Kreise*) was established. Subsequently, a range of Ministries was created for all branches of the economy, which were directly controlled by the Party.

A complex system of decentralized decision-making was replaced by a centralized, 'top down' system of control, based on the various branches of the economy. The communes were left without any clear legal competence, or any independent sources of finance. They became a shadow of their former selves, carrying out limited functions for the central Ministries, with money allocated for the tasks they were given. The emasculation of the local elected councils, and the absence of any local public participation, had disastrous consequences for the development of local communities. At the same time, town planners and architects were prevented from exercising their traditional functions of organizing urban development; they were confined to the implementation of plans for 'solving the housing problem' laid down by the Ministries. One of the consequences was an unjustified emphasis on the development of East Berlin and the central cities of the governmental districts; small towns and villages were badly neglected.

### Excessive building in concrete

When the Housing Programme was introduced in 1973, there was no effective housebuilding industry. However, in the post-war years experience had been gained in pulverizing the massive mounds of rubble from the destroyed cities for use in the production of concrete. The idea of using concrete components was taken up, and developed into a system using large concrete panels for housing construction. The pressure for success in mass-producing housing, at any cost, led the Party leadership to give overriding priority to the use of this system. A nation-wide industry was set up for producing standardized panels and assembling them for housing construction.

With this system, fast rates of construction were achieved; the construction of an apartment, from the laying of foundations to the handing over of the key, took only 700 man-hours, of which 290 were for the construction of the panels. Nor did the housing produced necessarily have to be monotonous over-large 'Russian-style' apartment blocks. Courageous architects and managers – often in the face of political criticism – showed that, used with imagination and sensitivity, this form of construction could produce housing which was varied and attractive to the public – and even acceptable to West German building firms.

Nevertheless, the proportion of housing built with this method (70–80 per cent) was far too high. Local styles of housing, adapted to the landscape and based on local materials, were largely lost. Thus the citizens' feeling of being at home in his community was undermined. The wish of many citizens for single-family housing was not sufficiently taken into account; this form of housing amounted to 10–11 per cent of new construction, which fell far below the potential demand.

## HOW TO SOLVE THE CURRENT PROBLEMS

The tasks of housing policy in the 'five new *Laender*' are:

(a) to mobilize the capital and resources needed for new and improved housing.

(b) to devise new, market-orientated forms of tenure and management.
(c) to re-establish the legitimacy of private ownership.
(d) to do the above without imposing excessive hardship on any group of citizens.
(e) to encourage people to stay put rather than move to West Germany.

Item (e) is in the interest of the 'old' Federal Republic, where the influx of East Germans has aggravated housing shortages, as well as of the 'five new *Laender*', which are losing highly qualified young people. The emigration is mainly in search of employment, but housing policy can play a role; if people have relatively cheap housing, they will be less inclined to move. These tasks require a series of radical changes in political and housing institutions.

### Federal, devolved government; privatization and democratization

A turning-away from central planning and a return to a devolved, federal system of government, in conjunction with privatization and the introduction of a 'socially responsible market economy', is a pre-condition for economic recovery and the establishment of a free society. The awakening of individual initiative and individual responsibility lies at the root of the reorganization of our country. Similarly, the restoration of autonomy for our *Laender* and communes, and their provision with independent financial and material resources, is an essential requirement for urban renewal and an improvement in the housing situation. It was in this spirit that the People's Assembly passed the Communal Property Act of 6 July 1990. Under this Act, the 'people's property', in the sense of housing and land which was vested in the nationalized enterprises, was transferred to the communes. The 317 KWV (*Kommunale Wohnungsverwaltungen*, or 'People's housing organizations) which administer around 2.8 million dwellings, will generally be transformed into limited companies, of which the main, or sole, shareholder is the commune. The local council can then decide whether it will retain ownership, or transfer ownership in whole or part to housing associations, cooperatives or individuals. The limited companies are supervised by a Supervisory Board, on which Tenants Associations and citizens groups are represented. It is foreseen that the Supervisory Boards will discuss the issues of rent levels and modernization, as well as questions of cleanliness and safety, taking the views of tenants into account.

The Act of July 1990 also gave the commune the right to nominate tenants for any dwelling which has been built or subsidized with public funds. For such dwellings, any tenancy agreement will be dependent on the tenant possessing an 'entitlement certificate' which the commune will distribute, on the basis of housing need, to people who seek housing in the area. Any person can apply for such a certificate if he and his family does not have 'appropriate' housing. 'Appropriateness' is defined in terms of floor-space, as follows: for 1 person, 45m$^2$; for 2 persons, 60m$^2$; for 3 persons, 3 rooms or 75m$^2$. The commune will establish priorities for the distribution of certificates.

In order to underpin the autonomy of the communes, it is essential that they should as soon as possible receive income from the Business Tax and Income Tax, as well as the other taxes and fees which finance local government in the Federal Republic. It is also important that they quickly build up competent departments dealing with housing, land transfers, and related fields.

New rules will also be needed for the 871 existing housing cooperatives, which manage around 1 million dwellings. According to recent legislation, they can acquire ownership of the land on which their buildings stand (which was previously 'people's property'), as long as there are no legal obstacles. In the longer term, the cooperatives will have to charge rents which cover all the capital and running costs.

### Prudent housing and rent policies

Under the legislation of July 1990, the existing fixed rents for all housing were retained. It was accepted that a change in rent levels would be undertaken in the light of the trends in income, and when housing allowances (*Wohngeld*) had been introduced, which will provide social support for tenants on low incomes.

Berlin presents a particular problem for housing and rent policy. With the growing together of the two parts of the city, a rapid equalization of incomes is to be expected, which will hopefully be carried out in a socially tolerable way. Only then will it be possible to begin a gradual equalization of rents, with the aim of introducing a housing market, with its potential advantages for citizens. The responsible authorities will have to consult each other, with the aim of limiting extortionate rent rises, and other undesirable phenomena. The 50,000 private rented dwellings in East Berlin can now be let to people from West Berlin (where rents are much higher). For the moment, however, the East Berlin tenants enjoy fixed rents and security of tenure. The task for the immediate future is to find solutions which will ensure a gradual equalization of incomes and rents, with the minimum of social hardship.

The prime requirement for solving the housing problem is the construction of more (new or renovated) dwellings; only then can there be a housing market with tolerable rent levels. A high level of construction will also provide much needed employment. The revival of housebuilding should therefore be a priority of the communes.

### The development of the construction and housing market

There must be a fundamental change in the structure of the building industry. The large construction combines must be dismantled, and replaced by a range of middling and small firms, in line with the industrial structure in the Federal Republic and other Western countries. At present, the larger cooperatives also have teams of maintenance workers attached to them; these should be disbanded and turned into independent contracting firms. The aim is to develop a more decentralized and flexible structure, which meets the needs of clients. The communes will play an increased role, even though they will not normally build housing.

### Housing needs

The need for housing is great. Forecasts for the period 1991 to 2000 arrive at the following total requirements: 600,000 to 700,000 new dwellings, to replace those which must be demolished because of their poor condition, and a further 200,000 to 300,000 new dwellings, to cater for increased demand, giving a total of 800,000 to 1 million over ten years. As a result of a demand which was previously frustrated, the number of single-family houses built each year is likely to rise from around 10,000 to two or three times this level. Incentives for saving with building societies can help people to realize their ambition to become home-owners.

A large part of the 60,000–70,000 dwellings needed each year for the replacement of unsound older housing will have to consist of subsidized 'social housing', since most of the inhabitants of East Germany are unlikely for many years to have the income or capital needed to buy housing or pay commercial rents. Savings per head are at present on average not more than 6,000 to 7,000 DM, and are unlikely to rise substantially until the late 1990s. In order to ensure good living conditions for all the people in the Eastern part of

Germany, substantial financial assistance from the Federal and *Laender* Governments will be needed for some time to come.

## THE PROBLEMS OF TRANSITION

The collapse of the communist state left an administrative vacuum, which is only slowly being filled. One widely reported episode in Berlin illustrates the chaotic situation with which the authorities have had to deal.

### The squatting riot in Berlin in November 1990

In November 1990, almost a year after peaceful demonstrations brought about a political revolution, Berlin saw a violent confrontation between police and squatters. Some 4,000 police and 400 squatters were involved. The episode was at least as much a political demonstration as a product of a 'housing shortage'. The squatters included substantial elements from the terrorist fringe (Red Army Faction) which used extreme brutality in a fight which they had already lost in West Germany. However, housing conditions are still so different in the two parts of Germany that the squatting must be assessed differently, even if the motivation of the predominantly young people involved may in both cases be that expressed by Berthold Brecht:

> Given that the houses stand,
> While you leave us homeless,
> We are going to take a hand,
> Since our holes don't suit us.

In the 'old' Federal Republic, landlords generally precipitated squatting incidents by emptying properties, with a view to rebuilding or modernization. (The squatting incidents in Berlin in 1980–2 followed insensitive behaviour by *Neue Heimat*, which left whole streets empty.) In the GDR, on the other hand, it was the inability of the public housing administration to maintain and manage its housing stock which forced tenants to take action. A combination of circumstances – extremely low rents, the lack of legal means to enforce the payment of even these low rents, ineffective maintenance and management systems – led to a situation in which an increasing number of dwellings became hard to let, suffered deterioration, and eventually had to be abandoned. In the final days of the GDR, the housing authorities did not even have reliable information on which dwellings were occupied and who the occupants were.

Under these circumstances, it became a common practice for empty run-down dwellings to be simply taken over by persons seeking housing. It was sometimes then possible for the occupiers to reach agreement with the housing authorities for the renovation of the dwellings; the authorities were usually willing to subsidize such renovation, if the occupiers were in a position to arrange for it to be undertaken 'unofficially'. In spite of this 'private enterprise', the housing policy followed in the GDR has led to a situation in which, in East Berlin in 1990, 25,000 dwellings were empty, while in the district in which the troubles occurred (Friedrichshain) 80 per cent of dwellings were in need of renovation. At the same time, this district had 4,000 people seeking housing, some of them for many years.

This background explains why, after the breakdown of state authority in November 1989, organized squatting in entire apartment blocks began to take place. When the newly-elected City Council began work on 5 June 1990, 71 apartment blocks were affected. The Council, operating in a confused political situation, decided to legitimize all squatting before 24 July, but to ban it thereafter. In assessing the correct policy to be adopted, it is necessary to distinguish between occupiers who are concerned to take up permanent residence, and improve the properties, and occupiers who present a threat to the security of their neighbours. (The ending of the authoritarian regime has had as one of its negative aspects the emergence of various small extremist groups.) It is nevertheless encouraging that, in a quarter of the cases, agreements have been reached for the occupiers to enter into tenancy agreements. Violent incidents like that in the Mainzerstrasse hit the headlines – and are symptomatic of the growing pains of the 'five new *Laender*' – but they affect only an insignificant percentage of the housing stock.

### The Unification Treaty

The legal framework for housing policy in the former GDR was laid down in the unification treaty. The outcome was (to quote a Federal Ministry booklet) that

> From 3 October 1990, and with only a few exceptions for a limited transitional period, the entire body of law on housing, rents and building in the West German *Laender* will apply in the *Laender*,

Brandenburg, Mecklenburg-Vorpommern, Sachsen, Sachsen-Anhalt and Thureringen, as well as in East Berlin. The basis for a general improvement in housing and town planning has thus been created.

The Ministry's optimistic conclusion may well, in general, be justified, but there are three problems arising from this imposition of West German law. One is that the 'old' Federal Republic abolished the status of 'non-profit housing associations' just before the GDR collapsed. In the optimistic period after the fall of the Wall, many people in the GDR hoped that the state housing companies could be turned into non-profit housing associations, but that is not now – in strict legal terms – possible.

Two further difficulties arise from specific provisions of the unification treaty. One is the provision that all *émigré* property compulsorily acquired since 1949 has to be given back to its original owners or their heirs. The final, democratic, Government of the GDR pleaded for the application, in most cases, of a 'financial compensation' rather than a 'giving back' principle – on the grounds that universal 'giving back' would provide work for the lawyers for years, while inhibiting the development or improvement of the property concerned, and causing hardship to existing occupiers.

The 'giving back' principle was, however, laid down. In the case of housing (although not industry), the issue has temporarily been put on one side by two legal provisions. First, ownership does not affect rent law: rents are still controlled, and are being raised in stages. Second, the provision that tenants may be given notice if the owner requires the dwelling for his own use will not be introduced before the end of 1992.

### The communes housing debts

A further problem arises from the provision in the treaty that the housing owned by state enterprises is to be handed over to the communes, together with existing debts. The outcome is that the communes now own a large stock of predominantly high-rise, system-built housing, with debts of around 85 per cent of its historic cost. It is widely accepted that most of this housing should be administered by companies which are legally independent of communes, even if the communes are the sole, or main, shareholder. The model is the former non-profit companies owned by West Germany municipalities. (This solution has been advocated by the *Gesamtverband der Wohnungswirtschaft ev*, – National Association of Housing Enterprises –

which has been formed by the former non-profit associations in West Germany, together with the former 'housing cooperatives' of the GDR.) There is, in addition, some scope for selling dwellings to tenants, and the Federal Government is giving subsidies to this end. However, although owner-occupation in general needs to be expanded, a cautious approach is advisable in the case of the large system-built housing blocks. These flats could well prove a trap for purchasers because of the uncertain costs of renovation. Conversely, it would be undesirable for communes to sell off the 'desirable residences', leaving themselves with the 'hard to let' housing; this would be a sure road to 'ghettoization'.

The council's current high level of debt, combined with the need to raise rents which at present do not even cover maintenance costs, creates a dilemma. Some communes, hard pressed for finance, may be inclined to sell off their housing merely as a means of improving their cash flow, irrespective of the longer-term implications for housing management. If, however, the housing is hived off to separate management companies, these companies will be forced to raise rents to a level which would create hardship for many tenants. The companies (and cooperatives) might, in fact, not be able to raise rents sufficiently, and some might be forced into bankruptcy. This would create enormous disruption.

In this situation, the National Association of Housing Enterprises has urged the Federal Government to remit the debts of the housing companies and cooperatives in East Germany. The long-term aim, it accepts, should be to achieve a gradual equalization of rent burdens in both parts of Germany, with the housing companies and cooperatives in the East earning a return on capital, as well as covering maintenance and repair costs. However, the present level of debt is an intolerable burden at a time of economic collapse, and is based on accounting valuations which (taking into account the cost of repairs) may often exceed the market value of the housing. A debt-forgiveness programme – the Association argues – would moderate the rise in housing costs, and so encourage people to stay put; it would thus be one of the most cost-effective forms of assistance to the five new *Laender*.

The National Association also proposes that the housing owned by the former 'cooperatives' in the GDR (17 per cent of the stock) should be transferred to cooperatives reformed under existing Federal law, and freed from debt. The cooperatives in the GDR were not cooperatives in the Western sense, but they did involve a degree of financial and managerial participation by tenants. The cooperatives of the 'old' Federal Republic were not affected by the law abolishing non-profit companies. It is too soon to say whether, in the end, the whole 'cooperative' stock can satisfactorily be managed in this way, but there is certainly a case for experimenting with cooperatives; a variety of tenures offers the best chance of meeting housing needs, given the limited resources available.

## The introduction of service charges

One of the consequences of the introduction of West German legislation on rents is the separation of 'rent' and service charges – for refuse collection, street cleaning, heating, hot water, and water supply. Wherever possible – e.g. for heating – individual controls and metering will be gradually introduced. Under the old regime, no separate charges were made for these services, and there was often no means of controlling the temperature in individual dwellings – except by opening the windows! The result was that the potential advantages of schemes such as district heating were squandered by wasteful use of energy. Individual charging and metering is to be welcomed as an encouragement to the rational use of resources – although the social security system will have to provide a safety net for households on low incomes.

## Renovation and affordability

In dealing with dilapidated and unmodernized housing, it would be a mistake to adopt the extremes of either 'knock it all down' or 'keep every dilapidated house'. Some housing has deteriorated to such an extent that the cost of renovation would exceed the cost of new building. In such cases, it would be better to demolish the buildings and replace them with good new construction, sensitive to its surroundings – except for buildings of special historical interest. But most housing could be renovated and modernized for much less than the cost of new building; modernization would also cause less social disruption. Moreover, some renovation can be organized or carried out by the occupiers themselves; this 'informal sector' needs to be used to the full, given the prevailing low level of savings and the shortage of capital in the region. The best mix of renovation and redevelopment can be decided only after studies on a 'building by building' basis. Planning surveys designed to improve not only the physical structure, but also to give more life and variety to monolithic housing estates, are under way.

Another issue concerns the level of renovation. For a given expenditure, it may be possible to bring three dwellings up to a moderate standard or one to a de luxe standard, and in current conditions the former course of action is preferable. However, the building regulations of the Federal Republic set some of the highest standards in the world, and there is a danger that the standards being set are too high. As a transitional measure, some dispensation to the *Laender* to relax national standards seems to be called for.

Given current income levels, much new building and rehabilitation will have to be subsidized, but a large part will have to be paid for by households. The current expenditure on housing as a percentage of income is 4 per cent in the new *Laender* and 19 per cent in the 'old' FRG; it is clear that households' expenditure will rise considerably. In some cases, this increased expenditure will be matched by higher incomes, in other cases it will not. Hence the need for a safety net provided by *Wohngeld* (housing allowance) and the social security system. The situation will have to be carefully monitored to ensure that the much needed improvement in housing conditions does not bear too heavily on the budgets of the more vulnerable households.


## POSTSCRIPT, DECEMBER 1991

The fears expressed at the time of unification – by citizens of the GDR who had no liking for communism – about the consequences of such an unprecedented change in the economic system have proved only too justified. Unemployment is now effectively over 20 per cent, and higher in some districts; many unemployed or pensioner families are facing financial pressures arising from the rise in housing and other costs; crime, political extremism and drugs have appeared on the scene.

All rents were raised by 1DM per square metre on 1 October 1991. The effects of this change (and of the introduction of service charges) are shown in this example for a 68 square metre flat in the settlement of Berlin-Marzahn. The rent will rise from 85 to 166 DM (per month); in West Germany, the rent would be at least 400 DM.

| | |
|---|---|
| Previous rent | 85 DM |
| From 1 October 1991 | |
| 10% reduction for service charges | −8 DM |
| Increase of 1 DM per sq. metre | +68 |
| Hot water and heating | +10 |
| Premium for location | +10 |
| New rent | 166 DM |

A survey of rents throughout the five new *Laender* suggests that rents-plus-service charges have risen from 83 to 295 DM a month. Service charges, in particular, are becoming a problem for some households. The regulations concerning their level are still unclear, and 'cost covering' charges could be onerously high, as a result of inefficient supply systems and poor thermal insulation. Energy conservation and efficiency in service provision need to be a priority.

There are signs of progress in small-scale repairs and improvements. With the establishment of small private firms, it is now possible to engage plumbers or heating engineers or roofers – something that, under the old regime, often involved long delays or devious black-market arrangements. There is a noticeable upsurge in the installation of new heating systems, etc. On the other hand, progress in structural improvements has been slow. Soft loans are available from the *Kreditanstalt fuer Wiederaufbau* (Finance Corporation for Reconstruction) but owners have to put up substantial amounts of capital, and neither communes nor individuals have much capital. (The ratio of personal capital per head in West and East Germany is roughly 15:1.) There are also uncertainties over ownership, and administrative bottlenecks. The tax concessions available to private owners are of little value to most resident landlords because of their low incomes; the main benefit will go to West German investors.

At the time of unification, some 42 per cent of housing was in state ownership. Claims have now been made by individuals (mainly in West Germany) for the return of 1 million dwellings, around 15 per cent of the housing stock; thus the housing owned by communes will before long be reduced to, at most, 27 per cent.

The lists of applicants for housing – especially from pensioners, handicapped people, and low-income families with children – are increasing, and communes are having difficulty in meeting the demand. One survey indicated that, in ten months up to June 1991, only 28 per cent of priority cases were provided with housing. At the moment, because most of the housing owned by communes has not been registered as 'social housing', there are no income limits on applicants, which is unfortunate. The housing allowance (*Wohngeld*) has been introduced but the take-up, at some 1.5 million out of an estimated 4 million eligible households, has so far been low.

The provision under which the new owners of confiscated housing can evict tenants if they require the dwelling for their own use, or that of close relatives, will take effect on 31 December 1992, but legal harassment of tenants began soon after unification. Many households fear the loss of their homes and two elderly East Berlin residents, threatened with eviction, recently committed suicide. Confrontation over contested housing has done more to alienate '*Ossis*' and '*Wessis*' than any other issue.

The problem of *Altlasten* (burdens from the past, in the form of unclear ownership, the costs of clearance, and pollution) threaten to hold up the redevelopment of redundant commercial, industrial or military sites for residential, and new commercial, uses. There is a danger that, when housebuilding gets under way, such sites will be neglected in favour of excessive green-field development.

These problems will be tackled more effectively as *Laender* and communes become better established but it can be seen that the progress of the former 'workers' and peasants' state' towards a 'socially responsible market economy' faces many difficulties.

# 6 Housing affordability in France

*Jean-Pierre Schaefer*

## HOUSING CONDITIONS IN GENERAL

Let us begin with the good news. Housing conditions in general have improved dramatically over the past twenty years, as a result of very high levels of construction and extensive renovation. Since 1970, the percentage of dwellings with a toilet and bathroom has risen from 56 to 90 per cent, while the percentage of dwellings that are statutorily 'overcrowded' has fallen from 24 to 12 per cent; the percentage that is seriously overcrowded is only 2 per cent.

Dwellings are classified in the survey according to 'comfort', as measured by sanitary facilities and central heating. On this definition, 90 per cent of all dwellings in 1988 were 'comfortable' or 'very comfortable'. The uncomfortable dwellings were mainly those built before 1914, including both tenanted and owner-occupied dwellings. Only 2 per cent of dwellings were 'seriously overcrowded' in 1988, although 10 per cent were 'slightly overcrowded'. At the other end of the spectrum, 48 per cent were 'slightly or moderately under-occupied' and 16 per cent 'highly under-occupied'.

### Housing satisfaction

According to opinion surveys, French households are predominantly satisfied, or very satisfied, with their housing, and the percentages in this category have risen over the past ten years. In 1988, only 6 per cent were 'dissatisfied', and only 3 per cent 'very dissatisfied'. Most of the 9 per cent in the 'dissatisfied' or 'very dissatisfied' categories were in housing classified (on the basis of overcrowding and 'comfort') as poor or bad. On the other hand, most people in these categories of housing were 'satisfied' or 'very satisfied'.

In interpreting these results, it must be borne in mind that factors

*Table 6.1* Housing conditions 1988

(000s)

| | Comfort level | | | | |
|---|---|---|---|---|---|
| | *1* | *2* | *3* | *4* | *Total* |
| Very overcrowded | 62 | 24 | 62 | 161 | 309 |
| Slightly overcrowded | 249 | 146 | 367 | 1,313 | 2,075 |
| Normal occupancy | 229 | 267 | 799 | 3,675 | 4,970 |
| Slightly under-occupied | 246 | 239 | 910 | 4,422 | 5,817 |
| Moderately under-occupied | 138 | 162 | 592 | 3,277 | 4,169 |
| Very under-occupied | 72 | 130 | 461 | 2,697 | 3,360 |
| Total | 996 | 968 | 3,191 | 15,545 | 20,700 |

1 Very uncomfortable: without bath or internal WC
2 Uncomfortable: without bath or internal WC
3 Comfortable: with bath/WC, without central heating
4 With central heating and sanitary facilities

*Source:* Ènquête Logement 1988 in M. Eenschoten, *Economie et Statistiques*, 24:61

other than the standard of the dwellings – such as the external physical and social environment – play a large part in satisfaction or dissatisfaction. Some people may be very satisfied with older housing in familiar surroundings, even though the dwellings lack modern amenities. On the other hand, some people in the less successful large housing estates may be dissatisfied because of the social problems of the district, even though the technical quality of the dwellings in the social rented sector is almost uniformly good. Younger people are more demanding, and hence more likely to be dissatisfied, than older people.

The opinion surveys also reveal that, if they had the opportunity, 65 per cent of households would like to live in a one-family house. This desire is tempered by the cost of access to owner-occupation; many people are unwilling to enter into the long-term financial commitment involved.

## THE PROBLEMS THAT REMAIN

In spite of the general improvement in housing conditions, problems remain, and for some social groups have become more severe. These problems fall into three main categories: housing affordability, homelessness and near-homelessness, and 'problem housing' of various types.

Table 6.2 Rent/income ratios 1978–88

|  | 1978 % | 1984 % | 1988 % |
|---|---|---|---|
| without charges, with AL/APL |  |  |  |
| HLM tenants | 6.8 | 7.5 | 9 |
| Private tenants | 11.5 | 13.1 | 15.1 |
| All tenants | 9.1 | 10.2 | 12.6 |
| with charges, without AL/APL |  |  |  |
| HLM tenants |  |  | 15 |
| Private tenants |  |  | 18 |
| with AL/APL |  |  |  |
| HLM tenants |  |  | 13 |
| Private tenants |  |  | 15 |

Source: Curci and Taffin (1991); Curci (1989)

### Affordability

Housing costs have tended to rise in relation to income. The average of rents and service charges for all tenants was estimated to be 16.4 per cent in 1988, compared with 13.6 per cent in 1984, and less than 5 per cent in the 1950s (at the peak of the housing crisis!) (Curci and Taffin 1991). This 16.4 per cent is after payment of the housing allowance (AL/APL), which is received by one-third of tenants (45 per cent in the 'social' HLM sector, 25 per cent in the private sector).

The rise in rent/income ratios to some extent reflects an improved quality of housing, but not all tenants are able or willing to pay for higher quality. The estimated ratios (including Housing Allowance, but without charges) for HLM and private tenants since 1978 are given in Table 6.2, which shows the steady rise in both sectors. Rents tend to be higher for new tenants, as compared to sitting tenants, and for new HLM dwellings as compared to older ones.

According to surveys carried out by UNFOHLM (Federation of HLM bodies) and the Housing Ministry, the average rent per month for subsidized dwellings at the end of 1989 was 1,100 FFR, varying according to the date of construction and method of financing. Rents ranged from 700 FFR in some older units to 2,000 FFR in units built since 1978 or recently renovated.

In the private sector, rents vary widely according to the local housing market. Rents in Paris itself are 45 per cent higher than in the outer Paris region, and double those in other cities. Typical monthly rents in 1988 were: 2,390 FFR for a 43 m$^2$ flat in Paris; 1,390 FFR for a 68 m$^2$ flat in a provincial city; 1,170 FFR for a 77 m$^2$ rural dwelling. Rents also vary according to the length of occupation by

*Table 6.3* Average rents per month 1988

|  | per dwelling FFR | per m² FFR |
|---|---|---|
| HLM sector | 1089 | 15.6 |
| Private sector, average | 1690 | 25 |
|   for new tenants | 1861 | 28.2 |
| Private sector (pre-1948) | 1189 | 18.2 |

*Source:* Curci, *Economie et Statistiques* no. 240

tenants. There are limits to annual increases in rents, and tenants who have been in occupation for many years tend to pay less than new tenants.

It must be stressed that, although a new social rented flat would cost the fairly high rent of 28 FFR per square metre per month, which might be above the rent in the private sector for older dwellings in many middle-sized towns, the tenant would be entitled to a personal housing allowance which, according to his/her income, could cover up to half of the cost. Apart from the rent, the tenant has to pay service charges covering water, cleaning, lift maintenance and (usually) heating. For houses or flats with individual heating, the gas or electricity bill is paid directly to the gas or electricity company.

Households on low incomes – especially those with no member in regular employment – face particularly severe affordability problems. Figure 6.1 illustrates the rise in the rent/income ratio as income falls. It also shows that the housing allowance is effective in limiting the rent/income ratio to about 17 per cent; however, not all households receive the allowance.

### Housing costs for owner-occupiers

Problems of affordability are not confined to the rented sector. Home-ownership rose sharply from the end of the 1960s onwards and, by 1988, 54 per cent of all households were owner-occupiers. Many loans from the 1981/5 period were taken out at a high and fixed rate of interest. This proved a severe burden for lower- or middle-income borrowers, but many of the loans have been renegotiated, with the assistance of public funds provided through the *Credit Foncier*; this is a financial body, under public control, engaged in loans for real estate. It is now possible to take out either variable-rate or fixed-rate mortgages. The latter have regained their attraction because of the low and stable inflation rate, with interest rates at 9–10 per cent. At the same time, the 'Law Scrivener' has discouraged

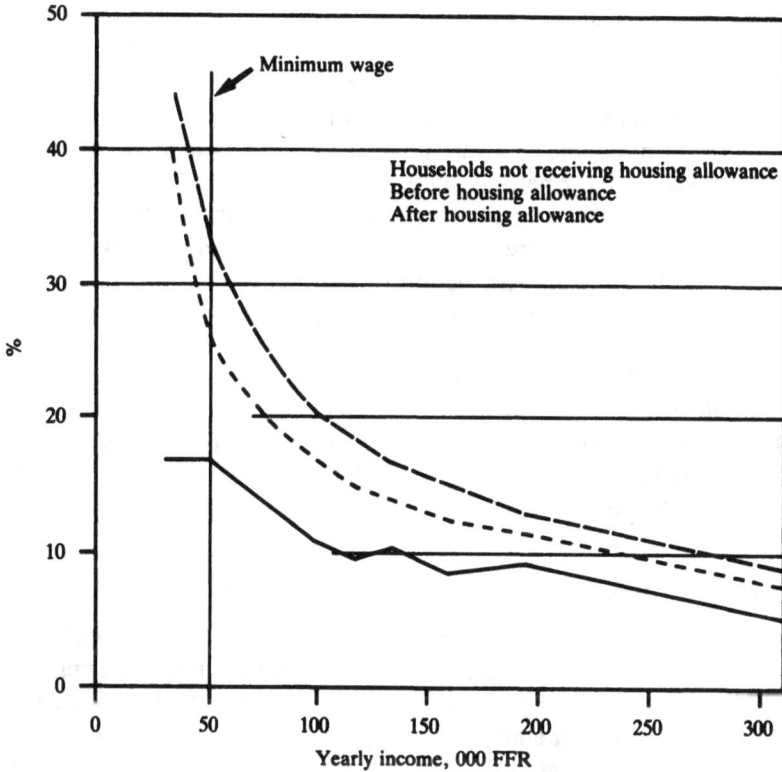

*Figure 6.1*   Rent*/income ratios 1988 (* including charges)
*Source*:   Housing Survey in G. Curci, *Economie et Statistiques*, 250:35

the granting of excessive loans to low-income households. In consequence, repossessions are not a serious problem at the present
time. Nevertheless, any further increase in home-ownership needs to
be considered very carefully, lest it increase the incidence of loan
default and repossession.

### Homelessness

The number of homeless people has been estimated at between
200,000 and 400,000; 0.4 to 0.8 per cent of the total population
(Brezinksi 1987). This includes squatters, people living in trailers,
precarious lodgings or hostels etc. However, all estimates of the

number of homeless people are subject to large ranges of error. There is no reliable statistical information on changes in the number of homeless people but there is evidence that, in the 1980s, a wider range of social groups has come to be affected, including young people and couples with children. A recent film, *Les Amants du Pont Neuf*, depicts the life of young, unemployed homeless people.

The various charitable agencies provide indications of the substantial number of people in severe poverty. *Secours Catholique* received 600,000 requests for help in one year, involving 2 million people, of whom one-fifth had a housing problem ranging from true homelessness to precarious tenure. The '*restaurants du Coeur*', which provide cheap meals for homeless people, have 365,000 people registered with them. Of these, 33 per cent are single people, 20 per cent are one-parent families, and 18 per cent are families with three or more children. The official category for homeless people is SDF (*Sans domicile fixe*, 'of no fixed address'). There are estimated to be 30,000 SDF persons in the Paris region, and 15,000 in Paris. A survey in December 1991, found 1,800 young SDF in the Paris Metro, and 200 at Roissy Airport.

### Problem housing

A final category of problems concerns the physical or social problems of certain categories of housing. On the one hand, there are a number of elderly owner-occupiers living in older, unmodernized housing, which they cannot maintain properly. On the other hand, there are the *grands ensembles* – the monolithic social housing built in the 1960s and early 1970s. Although the actual quality of the flats is good, the estates still suffer from the lack of an urban ambience. Moreover, racial tension has emerged as they have developed a relatively large proportion of foreign and unemployed people, with a low proportion of middle-income households. There were riots with racial undertones in the summer of 1991, even in estates where there had been considerable expenditure on social facilities.

## THE CAUSES OF THESE PROBLEMS

### Demography and income distribution

Many interrelated causal factors are involved in the problems outlined above: they include changes in household numbers and size; changes in the distribution of income; and changes in rental policy.

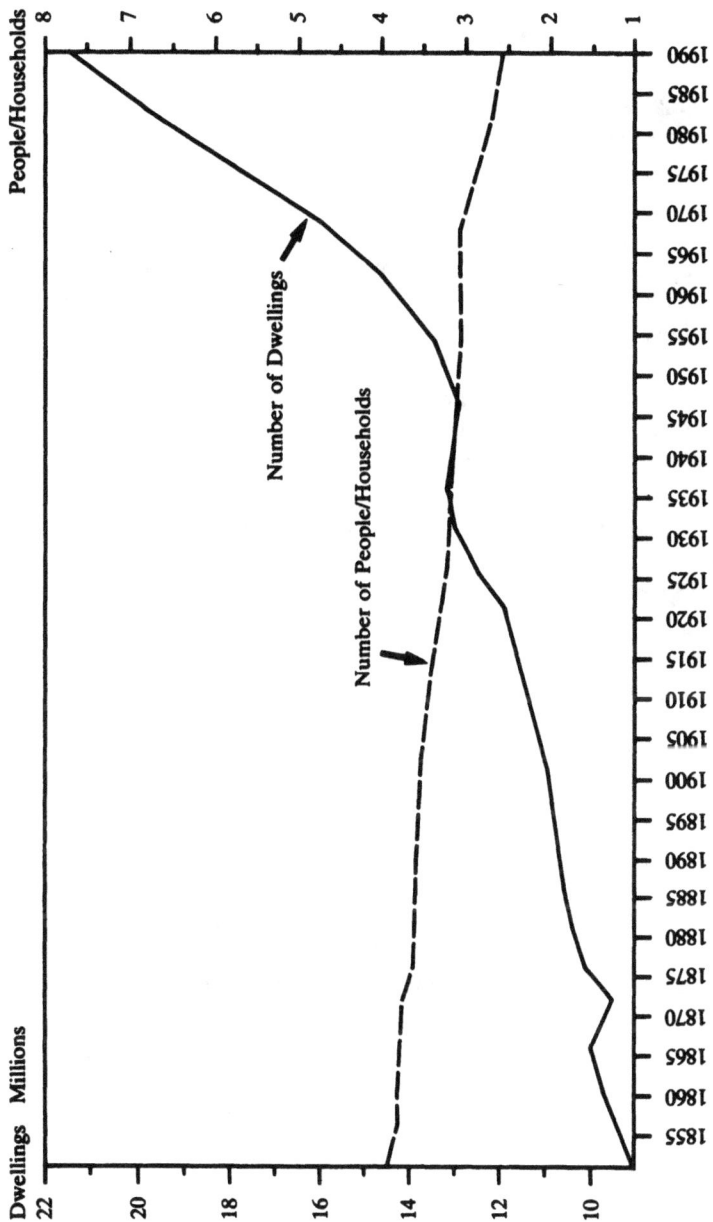

*Figure 6.2* Number of dwellings and average household size 1900–90
*Source:* C. Taffin *Economie et Statistiques*, 240:82

Between the 1950s and the 1970s, France experienced unprecedented high rates of population growth and urbanization. This increased demand was met by high levels of construction, under a system characterized by extensive state intervention. Both population growth and urbanization declined in the 1980s, but the number of households continued to increase, for several reasons: young people leave home earlier; the divorce rate and the number of single-parent families have risen; there has been a rise in life expectancy and a fall in family size. The effect of these social changes has been a steady decline in the average size of household, from 3 in 1975 to just over 2.5 today (Figure 6.2). This decline in household size tends to increase demand for housing.

At the same time, economic changes have brought about an increase in relative poverty, and so increased the problems of housing affordability for poorer households. During the 1980s, France achieved price stability comparable to that in Germany and the Netherlands; unfortunately, it was accompanied by an unemployment rate of around 9 per cent (much higher among young immigrants) and an increase in relative poverty. Average household income (in terms of purchasing power) has stagnated, and the reduction in inequality which took place in the 1970s, fuelled by economic growth, has stopped. Statistics are, however, available only up to 1984. In 1984 the income of the poorest two deciles of households, in constant francs, was slightly lower than in 1979. However, the income of the highest decile had also fallen. The situation was thus not simply that the poor had got poorer and the rich richer, but that the middle classes had gained at the expense of both.

In the early 1980s there was a fall in real wages; between 1979 and 1984, the average wages of workers who were heads of families dropped by 6.3 per cent in constant francs. This loss of income was partly compensated by an increase in female employment, although this was rarely the case in families with three or more children. The position of the low-paid has been upheld by the minimum wage, which covers 20 per cent of the workforce, and a high level of transfer payments has been directed to those on low incomes. Nevertheless, the situation of the poor, and especially of large families on low incomes, has worsened. Of the 20.7 million households in France in 1988 (excluding overseas territories) over 3 million had incomes of less than the minimum wage of 4,600 FFR a month. Some 1.5 million were below the official poverty line of 55 FFR per head per day.

It is difficult to assess how far the distribution of income changed between 1985 and 1990. The emphasis on stock exchange earnings and tax avoidance has probably benefited higher income, but not lower income, groups. There is certainly a popular perception that inequality in incomes has increased, which is a serious challenge for the Socialist government.

Unlike some other groups of the population, however, the income of the elderly has not fallen; on the contrary, it has steadily improved. As a result of a rise in pensions and the relatively low number of people born between 1915 and 1930, the income of elderly people has risen to equal that of the population as a whole. Poverty and homelessness is mainly a problem of a (rather small) number of young and middle-aged people. Many elderly people live in 'uncomfortable' housing, but this is due less to a lack of money than to a reluctance to invest, to borrow, or to change their life-style.

### Changes in rent levels

Another aspect of affordability problems is the rise in rent levels. This is associated with the greater freedom in setting rents which has been given to both the private and the public sectors. The private rented sector, with some 20 per cent of the stock, is still an important sector, catering for households of all income groups. The Rent Act of 1948 freed the rents of new dwellings but continued the freeze on the rents of old 'uncomfortable' dwellings, built before 1948, which had been in force during the inter-war period – and which had contributed to considerable deterioration in the older housing stock. In recent years, the view has been increasingly accepted that landlords should be allowed to charge remunerative rents, and rents have been allowed to rise. When pre-1948 dwellings are modernized, the more relaxed general provisions for rent increases apply (see Figure 6.3). There has been a marked tendency, in the regions with tight housing markets, for pre-1948 housing to be modernized, with a consequent sharp increase in rents.

For housing built since 1948, the initial rent has not been controlled, although there have been various provisions regulating rent rises for sitting tenants. These provisions have oscillated between 'market orientated' policies and more rigid control. The current situation is based on the 'Law Mermaz-Malandain' 1989. For newly-built property, the rent is freely decided by the landlord. For a new tenant in a previously rented dwelling, the rent should be in accordance with 'the current level of rent in the neighbourhood'. For a tenant with a lease, or taking out a new lease (for between three and six years), the rent can be increased yearly in accordance with the

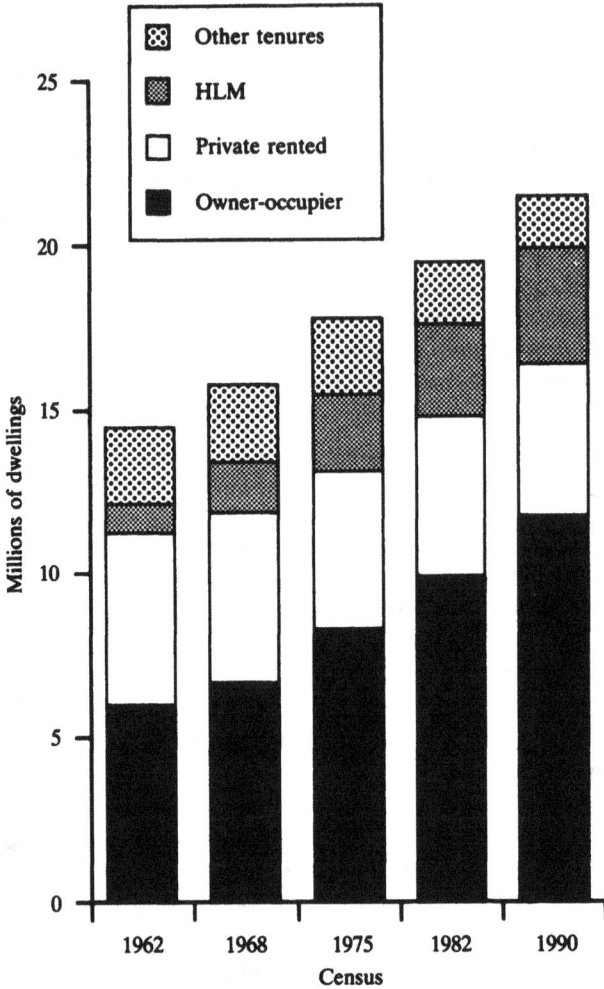

*Figure 6.3*   Housing stock, 1962–90
*Source:*   Censuses

index of building costs. However, rents may be temporarily frozen by the government in districts with disturbed market conditions; they were frozen in Paris in 1989 and 1990.

In order to measure the 'current level of rent in a neighbourhood' a rent table (*Observatoire des Loyers*), similar to the German *Mietspiegel*, has been produced for Paris, and others are planned for some major cities. It gives the typical rent (per square metre) classified by area and quality of dwelling. It is produced by a team of valuers, under the control of a committee including landlord and tenant associations, lawyers, and State and local authority representatives.

Thus, although there are still complex regulations concerning rent levels, they do not generally hold rents permanently below market levels. The relaxation of rent control in the 1980s is generally considered to have been necessary to maintain the condition of the housing stock, and to retain a viable private rented sector. It can, however, give rise to affordability problems for lower income tenants. In the more prosperous cities, where employment prospects are best, cheap unmodernized housing is now very rare. Such housing is increasingly confined to rural areas and some declining older industrial cities.

### The HLM sector

Current housing problems cannot be attributed to cut-backs in the supply of social (HLM) housing. On the contrary, the stock of HLM housing has steadily increased (Figure 6.3). The construction of HLM housing has fallen only slightly during the 1980s and there has been no general policy of selling off HLM housing. The social housing organizations are generally accepted as performing an indispensable social function, and they themselves wish to retain their 'good' dwellings. Sales are permitted when approved by the local authority and the State, but they have been confined to programmes designed to modernize large estates and diversify the pattern of tenure; they amount to less than 0.3 per cent of the stock each year.

Although there has been no 'attack' on social housing, the expenditure on 'bricks and mortar' subsidies has fallen, while the expenditure on the housing allowance has risen. The cut-back on 'bricks and mortar' subsidies has been accompanied by rises in the level of HLM rents; these have also resulted from the need for the HLM organizations to earn sufficient surpluses to undertake regular maintenance and improvement work.

### The housing stock

Current problems have to be seen against a background of a fundamental change in the housing situation. From the 1950s to the early 1970s, both the problem and the policy response were relatively simple – an overall housing shortage, and large scale, primarily State-sponsored, production programmes. In the 1980s, a much more complex situation emerged. The figures for the total housing stock might suggest that the supply of housing is adequate, but they need some qualification. The total housing stock rose from 18.3 million in 1968 to 26 million in 1990. Of this total, 2.7 million dwellings were second homes and 1.8 million were vacant, leaving 21.5 million 'principal residences'. The statisticians define the number of households as equal to the number of principal residences; there are also, of course, people living in hostels, hospitals etc., and homeless people.

The number of second homes and vacant dwellings has risen sharply since the 1960s. The former reflects a love affair of urban flat-dwellers with a 'house in the country' for *le weekend*. The high number of vacant homes has various causes. Most of these dwellings are 'uncomfortable' older housing in either traditional urban cores or in rural areas. The proprietors are too poor or too hesitant to invest in improvement; in some cases there is little demand for housing in the area; these units are therefore effectively outside the market. Other vacant housing arises from the normal turnover of the housing stock, as households leave for various reasons.

It is sometimes suggested that vacant HLM or private dwellings should be used for people seeking housing. The issue is unfortunately not so simple. The HLM has 100,000 vacant dwellings, but this is only 3 per cent of the sector's total stock. This is in line with the 2–3 per cent vacancy rate which is usually considered to be the result of normal turnover, and much lower than the rate of 8 per cent in the private rented sector. Even in the private rented sector, only a small proportion of the currently vacant units could realistically be brought into use at the present time, and to do so would involve a more active policy of urban renewal, including grants and loans for the owners.

The 21.5 million occupied principal homes are divided into four main tenures: owner-occupiers; the social rented housing (HLM), in which rents are subsidized and controlled; the private rented sector; and 'other tenures'. A small proportion of the private rented sector is still subject to rent control dating from the 1948 Act, but the rest is uncontrolled, apart from limitations on the amount by which rents may be raised in any one year. 'Other tenures' is a pot-pourri covering free housing, rented furnished dwellings, housing connected with a job, and some farm housing. In addition to 'normal' houses, there are (according to the 1990 Census) 19,000 dwellings, housing 40,000 people, which were classed as 'shanties' or 'provisional shelters'.

*Table 6.4* Housing tenure 1990

|  | thousands | % |
|---|---|---|
| Owner-occupier | 11,721 | 54 |
| Tenants   HLM sector | 3,133 | 24 |
|           private sector (free) | 4,593 | 22 |
|           private sector (pre-1948) | 0.50 | 2 |
| Other tenures | 1.588 | 7 |
| TOTAL | 21,535 | 100 |
| Second homes | 2,414 |  |
| Vacant | 1,896 |  |

*Source:* Housing Census 1990

There has been a marked rise in owner-occupation since 1975. The share of the social sector has also increased, whereas the private rented sector and 'other tenures' are slowly declining. The decline in private tenancy and 'other tenures' is concentrated in the stock of cheap, uncomfortable dwellings. This fall in the supply of cheap private housing has increased the demand for social rented housing from low-income households.

With more than 3 million dwellings, the social rented sector is similar in size to the social sector in West Germany, but proportionally smaller than that in the Netherlands or the United Kingdom. The sector is defined as covering dwellings whose construction is backed by financial aid from the State; this implies special regulations of the level of rents, rent increases, and the allocation of the dwellings. The dwellings may be let only to households whose income is below a fixed ceiling. In 1988, this ceiling gave theoretical access to 60 per cent of all French households.

**Housing output**

The number of housing units completed rose to very high levels in the 1960s and 1970s. It peaked at 516,000 in 1975. There was then a decline to 295,000 in 1985, followed by a slight recovery to over 300,000 (Table 6.5). Within this total, there has been a sharp reduction in the number of state-aided houses built for sale; building for sale has increasingly been handed over to the unsubsidized sector, although low-income buyers can receive the housing allowance. There has also been a decline in State-aided building for rent. However, since 1980 between 50,000 and 70,000 State-aided rental units have been built every year.

*Table 6.5* Housing units completed 1980, 1985, 1989
(000s)

|  | *1980* | *1985* | *1989* |
|---|---|---|---|
| State-aided rental (PLA) | 60 | 65 | 50 |
| State-aided owner-occupation (PAP) | 120 | 93 | 48 |
| PC* | 100 | 105 | 105 |
| Unsubsidized sector | 117 | 32 | 136 |
| TOTAL | 397 | 295 | 339 |

* PC, *prêt conventionné:* loan at a concessionary rate, without state aid but with personal aid if the owner is entitled to it.

*Source: Ministrie d'Equipment*, Housing Ministry

**Housing the poor**

People with low incomes, who face particularly severe affordability problems, are not concentrated in any one tenure; their distribution between the main tenure groups is in fact fairly 'normal'. It can be seen, however, from Table 6.6 that households on 'very low incomes' are under-represented among mortgagors and tenants of post-1948 housing, and over-represented in 'other tenures'. The same is true of the much larger group on 'low incomes'. This group is also over-represented among pre-1948 tenants and HLM tenants.

In 1984, the HLM sector had its 'normal' share of households on very low incomes, but the trend since 1984 has been upwards, as a result of higher unemployment among young people and a decline in the supply of cheap privately rented property. However, the 'residual-ization' of social rented housing is not (yet) as marked as in the UK, and there is no comparison with the 'ghettos' of the USA.

The slight over-representation of the very poor among outright owners reflects the substantial number of elderly people living in older, unmodernized housing which they cannot maintain properly, and which is often in a poor state of repair. This situation is causing growing concern, and indicates the limits to a policy of increasing owner-occupation.

**THE EVOLUTION OF HOUSING POLICY**

Housing policy in France over the past thirty years has been characterized by considerable continuity. There have been no 'U-turns', although policy has gradually been adapted to the evolving situation. In the decade 1953 to 1963, the main aim was to maximize

*Table 6.6* The distribution of low-income groups by tenure 1984

| No. of households | Very low income (1) 1,437,000 % | Low income (2) 7,527,000 % |
|---|---|---|
| Percentage of all households | 7 | 37 |
| Owner-occupiers, outright | 8 | 40 |
| Owner-occupiers, mortgagors | 3.5 | 28 |
| Tenants, private, pre-1948 | 8 | 49 |
| Tenants, private, post-1948 | 4.5 | 28 |
| Tenants, HLM | 7 | 49 |
| Other tenures | 12 | 49 |

(1) <20,000 FFR/year per capita; (2) <40,000 FFR/year per capita

*Note:* To use this table, each percentage should be compared with the 'percentage of all households' at the head of the column, to indicate under- or over-representation. Thus households with 'very low incomes', constituting 7 per cent of all households, are over-represented in 'other tenures' (12%), and proportionally represented in HLM dwellings (7%).

*Source:* Housing Survey 1984

new construction, with the emphasis on subsidized 'social housing' (Pearsall 1984). The following decade, 1963–73, saw a change of emphasis, with commercial enterprise and private finance being encouraged to play a larger role, and with households being forced to make a higher contribution to their housing costs. Many changes were made to this end, but in rather a piecemeal way. It was felt that a more comprehensive review was needed, and this was provided by two commissions (named Nora and Barre after their chairmen) which produced reports in 1975.

These reports criticized the existing housing subsidies as being too large in aggregate; poorly targeted, often helping the well-off at the expense of the needy; and over-centralized and inflexible. They recommended that the market mechanism should be relied on to provide most housing, but that the State should retain a substantial role, to assist the 'casualties' of a market system, and to intervene in cases (such as the rehabilitation of older housing) where the market system often did not work, or only at excessive social cost.

As a result of the discussion stimulated by the reports, the Housing Act of 1977 introduced several changes in housing finance:

(1) A new system of 'bricks and mortar' subsidies (PLA), linked to higher rents, but also to entitlement to
(2) a more generous system of housing allowance (APL), available to both tenants and first-time home-buyers.

(3) More assistance for rehabilitation, both for social (HLM) housing and private rented housing. The HLM housing became eligible for a new subsidy, PALULOS, while private rented housing became eligible for a new means-tested subsidy (PAH).
(4) Home-ownership was assisted by a new system of contractual loans (PC).

The 1977 Act was passed under the conservative Presidency of Giscard d'Estaing but its provisions have been for the most part continued under the socialist Presidency of François Mitterand.

The expansion of owner-occupation after 1977 was encouraged by two types of subsidized loans for owner-occupation: PAP and PC (*Prêt conventionné*, contractual loan). The PAP was heavily subsidized, and provided by the *Crédit Foncier de France*, a publicly-controlled bank. The PC was provided by private banks, and not subsidized, although it carried an entitlement to the housing allowance; it was designed mainly for middle-income groups. The PAP was a major factor in making owner-occupation possible for middle-income households, but has been cut back sharply since 1985. Thus a reasonable balance has been preserved between support for owner-occupation and other tenures.

### The 'one per cent'

A distinctively French housing subsidy is the employer's contribution to housing finance, or 'one per cent'. All private firms employing more than ten workers have to contribute (currently) 0.43 per cent of their wage bill to housing finance. They may invest it directly in subsidized housing for their employees, but more usually they provide funds to employers' organizations. These employers' organizations then make soft loans (at around 1 per cent interest) for the construction of dwellings for owner-occupation or renting; the loans are made directly to owner-occupiers or to HLM organizations. In the latter case, the employers' organizations enjoy allocation rights for their members' employees. As the money is a loan, the employers' associations are now receiving substantial repayments, which can again be lent out. This system has made a considerable contribution to providing affordable housing for many employees. It is, however, declining in importance – as indicated by the fact that the 'one per cent' is now only 0.43 per cent, and some of the money is now being used for the fund which provides the housing allowance. There is a possibility that the system may be attacked by the EC as being contrary to 'harmonization'.

To sum up, we can say that the effect of the complex French system of housing finance is to support a variety of tenures ranging from the purely 'social' to the purely private, and with various intermediate stages. Subsidies have been targeted reasonably well on middle- and lower-income households, irrespective of their type of tenure. The criticism can, however, be made that, in practice, the subsidies often do not benefit the very poorest households.

*Table 6.7* Subsidized loans for owner-occupation 1985, 1988
(Total numbers outstanding)

|  | *1985* | *1988* |
|---|---|---|
| Households getting a PAP | 122,000 | 81,000 |
| Households getting a PC | 195,000 | 180,000 |
| of which receiving |  |  |
| housing benefit | 82,000 | 56,000 |

*Source:* Ministrie d'Equipment

### The institutions of housing policy

There are a large number of public, or quasi-public bodies concerned with housing, which are usually referred to by their acronyms. The sharp divisions which prevail in some countries between 'public' and 'private' housing tend to be blurred in France by the existence of private bodies under public control and by public intervention in the supply of building land and credit.

As is well known, France once had a highly centralized administrative system, with the Prefects in the 95 Departments acting as agents for 'Paris'. Over the past twenty, and especially the past ten years, however, there has been extensive devolution of powers to local authorities (especially the cities and departments) and regional governments. Local authorities play an increasing role in analysing and dealing with their housing problems but they do not supply housing directly.

### Social housing

'Public sector' housing is represented by the HLM organizations (*Habitations à Loyer Modéré*, dwellings at moderate rents) which consist of nearly one thousand quasi-autonomous bodies, with an average stock of around 4,600 housing units. The HLM organizations fall into two main groups. On the one hand are 236 OPs (*Offices Publics d'HLM*, municipal agencies) and 20 OPACs (*Offices publics d'aménagement et construction*) which are sponsored by local authorities and regional governments respectively. They are set up under legislation covering the public sector. They are managed by a Board of Directors representing the local or regional authority, financial interests (especially savings banks), the tenants, and members nominated by the local prefect.

On the other hand are the 354 SAs (*Sociétés Anonymes*), which are non-profit limited liability companies. They are sponsored by private sector firms or associations, Chambers of Commerce, or public bodies mainly concerned with providing housing for their employees. They are entitled to subsidies in the same way as OP–OPACs, but they often also obtain money from 'the 1 per cent'. A distinctive organization in the SA group is the *Groupe SCIC* (Real Estate Branch of the *Caisse des Depots*. The latter is a large financial institution managing assets of nearly 1,600 billion FFR; it manages tax-exempt passbook savings, life insurance, mutual funds, and acts as a trustee for public or semi-public bodies. The Groupe SCIC manages 180,000 dwellings throughout France, with a network of companies (SA, HLM or SCI) of private status. It was originally a highly centralized Paris-based organization, but is now fairly decentralized, with a holding company (SCIC SA) based in Boulogne-Billancourt, near Paris, and various subsidiary companies dealing with the social rented sector, owner-occupation and real estate management.

The main 'bricks and mortar' subsidy for social rented housing consists of a grant (currently 12 per cent of the cost) and a soft loan for 80 per cent. The loan is known as the *Prêt locatif aidé* (PLA), 'assisted rental housing loan'. It is provided by the *Caisse des Depots*, which in turn derives much of its finance from the network of public savings banks. The rate is currently 5.8 per cent; although this is a cheap rate, it is substantially above the rates which were charged before 1977. There has therefore been a sharp increase in the rents of HLM dwellings financed under the new arrangements. If a dwelling is built with a PLA loan, the owner has to follow social sector regulations concerning rent, rent increases, allocations and income ceilings.

Rights to nominate tenants are shared between local authorities, the State, various partners, such as collectors of the 'one per cent', and the landlord organization itself. (The term 'State' in France means the centrally-organized system of public administration, represented in each Department by the Prefect.) There used to be considerable differences between allocation procedures for the housing agencies (OP–OPAC) and the companies (SA). The agencies allocate tenants through a committee which includes representatives of the local authority and of tenants. The companies were until recently free to organize their own allocation, and tended to attract tenants of a higher income level than the agencies. However, new regulations oblige them to set up an allocation committee like that used by the housing agencies; the need for negotiation with the local authorities about allocation will be emphasized.

The criteria for allocating tenants to social housing include, in practice, the ability to pay rent and charges, taking the housing allowance into account. There is a recommended, although not legal, limit of the rent-to-income ratio of 25 per cent. The system, at least until recently, has given a *de facto* priority to households with regular income; the unemployed, part-time employees and young people often find it difficult to obtain social housing. Once allocated a dwelling, a tenant may stay in it even if his/her income rises above the qualifying limit. In this case, however, a higher rent may be charged – although it would not be as high as in the private sector.

### Tenure and social composition

As middle-income households have increasingly become owner-occupiers, the private rented sector has come to have a higher proportion of low-income households. There has also been a tendency for more low-income people to apply for social rented units, because of their high quality and relatively low rents. This has begun to change the responsibilities of the HLM movement, although a high proportion of tenants are still 'traditional' in the sense of having regular wages, a moderate income level and a slow rate of turnover.

It is sometimes argued that the social sector has 'failed to house the poor' (Willmott and Murie 1988). This seems a little unfair. It is true that the system likes to have tenants who can pay rent regularly; regular rent receipts are essential for the financial viability of enterprises that do not enjoy open-ended state guarantees. Thus very poor people are more often found in tenures which do not involve regular payment, including outright owners of usually dilapidated properties and 'other tenures'.

However, the situation is changing. The HLM movement was created at the end of the nineteenth century to serve the needs of employees who could not afford home-ownership or the better type of private rented property. According to the Law Siegfried of 1884, the social sector was created to house 'persons who are not owner-occupiers, especially workers or clerks, living from their wages'. This principle was more or less maintained up to the 1960s, when social housing became a major component of the large building programme which was initiated. Social housing was targeted on both 'blue-' and 'white-collar' workers, who were all suffering from a housing crisis resulting from the low level of construction between 1914 and 1953. During the 1980s, the HLM sector came to have an increasing proportion of single-parent families, large families, immigrants and low-income households. Single-parent families made up 3 per cent of the total number of households in 1984 but 6.1 per cent of the households in the HLM sector (Table 6.8). The HLM sector had 15 per cent of the total housing stock in 1984, but housed two-thirds of the families with four or more children. In 1948, the HLM sector housed 48 per cent of households with incomes below the national average; in 1984, the figure was 59 per cent. At the same time, rents in the HLM sector have risen even more sharply than in the private sector (although they are, on average, well below the levels in the 'free' private sector) and tenants are increasingly dependent on the housing allowance. The HLM movement is having to develop a 'social work' aspect, and this presents it with a considerable challenge.

*Table 6.8* Social composition by tenure 1984

|  | % of all tenants in the tenure | | |
| --- | --- | --- | --- |
|  | 4+ children | single parents | foreigners |
| HLM | 6.1 | 11.2 | 12 |
| Private tenants | 1.6 | 7.8 | 8 |

*Source:* Housing Survey 1984

**The 'aided sector' and ANAH**

In addition to the various HLM organizations and 'the 1 per cent' there is also the 'aided sector', in which private firms are given subsidies for providing subsidized housing, for rent and for sale. This sector, which was expanded in the 1950s and 1960s, has been cut back sharply in recent years, as the emphasis has shifted to the housing allowance and tax rebates. The 1989 Housing Law, however, contains some provisions for assistance to private landlords who agree to house homeless people (see following paragraph).

There is also a levy/subsidy scheme designed to encourage rehabilitation in the private rented sector. In 1971, the *Agence Nationale pour l'Amélioration de l'Habitat* (ANAH, National Agency for Housing Improvement) was formed. This is a quasi-autonomous public body consisting of representatives of landlords, tenants and the State, financed by a grant from the State and a levy of 3 per cent on private rents. The money is distributed to landlords as grants towards the cost of specified improvements and repairs.

**Public expenditure on housing in the 1980s**

The 1977 Act was based, in part, on a belief that a switch of some public expenditure from social housing to the housing allowance would deal with housing needs more effectively. There was, however, no violent change, and no fall in total expenditure. 'Bricks and mortar' subsidies fell by about a quarter (in constant-value francs) while expenditure on the housing allowance nearly doubled. Total expenditure increased, at a time of fiscal austerity. Current housing problems cannot therefore be blamed on a lack of government spending.

*Table 6.9* Public expenditure on housing in 1980–90 prices (in billion FFR)

|  | *1980* | *1985* | *1990* |
|---|---|---|---|
| Bricks & mortar subsidies | 19.2 | 18.6 | 14.7 |
| Housing allowance, AL–APL | 12.5 | 20.4 | 23.3 |
| ANAH & 1% | 5.7 | 5.9 | 4.7 |
| Tax rebates | 14.6 | 12.5 | 14.4 |
| TOTAL | 52.0 | 57.4 | 57.1 |

*Source:* Ministry of Housing

**Future housing needs**

It has been estimated that, because of the continuing increase in the number of households, the shifts in population, and the need to replace dwellings which are demolished, 320–350,000 new dwellings a year will be needed during the rest of the 1990s. Although somewhat speculative, figures of this order tend to be accepted as non-binding targets by the authorities. The current level of construction, at just over 300,000, is somewhat below these figures, but not dramatically so; in terms of per 1,000 inhabitants, it is among the highest in Europe.

It is generally accepted that, within this total, the construction in the social rented sector should remain at around its present level of

50–60,000 units a year. There is no demand from the conservative political parties or the employers' organizations to cut the production of social housing, or to sell off the existing stock.

## THE 1989 'LAW FOR IMPLEMENTING A RIGHT TO HOUSING'

This Law is the most important piece of legislation since the 1977 Housing Law – which, among other things, signalled a switch from 'bricks and mortar' subsidies to personal housing aid. The 'right to housing' does not enable an individual to claim any particular type of housing from the authorities. On the other hand, it is more than a mere aspiration, in that it provides the justification for various specific programmes, and could conceivably be taken into account by the courts in decisions on these programmes.

The 1989 Law is based on three principles:

1 An attack on homelessness and bad housing requires a partnership of public and private bodies.
2 The supply of housing should be diversified in terms of tenure and type: private and social sector; new and old dwellings; large and small dwellings; flats and houses.
3 An attempt should be made to achieve a real integration of homeless people into society.

Starting from these principles, the Law provides the legislative basis for a number of programmes:

### 1 *Plan départemental d'action pour le logement des personnes defavorisées* (Departmental housing plan for under-privileged persons)

A scheme for the 'most underprivileged' is to be organized at each departmental level, with the involvement of local authorities. The schemes should take local housing markets, local levels of homelessness, and the social composition of homeless persons into account, and should cover matters wider than the provision of housing. The schemes should provide help for homeless or near-homeless people in the most cost-effective ways; these can include loans, guarantees to landlords, grants for reducing debts like rent arrears and electricity or gas bills. The aim should be to get households' finances on to a sustainable basis, and prevent the downward spiral which can be initiated by an accumulation of housing and other debt. A social welfare or charitable body should be made responsible for giving households counselling on debt management. The funds for these schemes are provided by the State, local and regional authorities, and local partners such as housing companies and welfare agencies.

### 2 Increasing the supply of private rented housing

Tax rebates are available for three years if housing is let to homeless persons at below a fixed rent level. There are provisions for a charity or non-profit association to lease the dwelling from the landlord and to be responsible for management and letting.

### 3 *Bail à réhabilitation* (Lease and rehabilitation)

There are provisions for soft loans to social housing companies or non-profit associations which take dilapidated property on a twelve year lease, rehabilitate it, and manage it according to social housing rules. The aim is to make use of vacant older property, and provide cheap rented dwellings in the urban cores. After twelve years, the dwelling will be returned to the landlord, or a new lease negotiated. The tenure will be designated 'social tenancy in privately owned dwellings'. This scheme is encountering some difficulties, because housing organizations are reluctant to buy isolated properties, and manage them in co-ownership with private partners.

### 4. Allocating dwellings for homeless people in the social sector (*Protocole d'occupation du patrimoine social, POPS*)

This is an attempt to prevent the formation of 'ghettos', when homeless people, immigrants and other 'problem' groups are concentrated in certain social sector organizations (notably the municipal agencies, rather than the companies), and in certain estates. The object is to spread such households throughout the social housing stock. Local authorities and housing bodies are to enter into an agreement establishing a 'balanced allocation'. In the absence of such an agreement, the State representative can directly allocate homeless households.

### 5 Increasing households' income

This programme is based on the philosophy that one way of meeting the housing needs of poor people is to enable them to pay for a 'normal' dwelling. It consists of two parts:

(a) Provisions to grant the housing allowance (APL) to a wider group of households, and also to enable it to be paid directly to the landlord, rather than the household. This is designed to overcome landlords' objections to letting to persons on the housing allowance, because of fear that they will be poor payers.

(b) *Revenu minimum d'insertion* (RMI, minimum income)
The provision for a minimum income, established in 1988, had been allocated to 400,000 persons by mid-1989. A payment is made to cover the difference between a ceiling income and the existing household income. In 1989, the ceiling income was 2,000 FFR for a single person, 1,000 FFR for a second member of the household, and 600 FFR for subsequent members. When the beneficiaries are owner-occupiers, or receive free housing, the payment is reduced. On the other hand, the beneficiaries are entitled to housing benefit; this will lead to 100,000 new beneficiaries. It is planned to link the payment of minimum income with an education and training programme, so that some of the recipients may be enabled to become independent.

6 Contributions from 'the one per cent'

An agreement has been reached between the state and the CIL (Association of 1 per cent collectors) to set aside, in 1990, a share of its funds (1,000 million FFR) for disadvantaged groups. The money can be used for building new dwellings, purchasing private rented property, renovating'existing units, providing low-cost home-ownership in overseas territories, or obtaining allocation rights for dwellings well suited to the needs of disadvantaged groups.

The intended beneficiaries of these measures are: the recipients of the minimum income; wage-earners in a precarious position (seasonal or part-time employees); wage-earners experiencing difficulties of access to housing (one-parent families, large families, immigrants). An additional 200 FFR will be used for social measures (advice, consultancy, experimental projects). This scheme is directed at social groups which are, albeit precariously, integrated into the economic system, and it seeks to involve industrial companies in the solution of the housing problems of these social groups. If the scheme proves successful, it may be repeated each year.

7 Administrative improvements

(a) Widening opportunities for low-income home-seekers.
The increasing demands on the HLM sector have made the allocation of needy tenants more difficult. At the same time, the relative newness of the HLM stock (approximately 90 per cent built since 1950), its high standards, and the limited opportunities for rent pooling, make the rents of new HLM housing relatively high – often higher than the free-market rents of unmodernized private dwellings. Attempts are therefore being made to, on the one hand, extend the range of housing available to low-income tenants and, on the other hand, to target housing benefit on those in greatest need. The housing allowance (APL) is being made available for a wider range of dwellings. But all the financial resources of the household are to be taken into account in deciding eligibility for benefit.

(b) Grants to the private rented sector
The grants provided by ANAH (*Agence National d'Améloriation de l'Habitat*) are to be increased if the housing is let to homeless or badly-housed households, or to associations acting on behalf of such people.

(c) Urban renewal and low-income households
OPAH (*Operation Programmée d'Amélioration de l'Habitat*) is a programme for the renewal of historic urban centres. It has now been supplemented by PST (*Programme Social Thématique*) which is targeted towards homeless or badly-housed households. It is undertaken by an agency, set up and financed by the local authority, which acts as an adviser to landlords and tenants on the technical, financial and social aspects of the programme.

8 The poor owner-occupier

Low-income owner-occupiers are entitled to PAH (*Prime a l'Amélioration de l'Habitat*). This is a grant of up to 24,000 FFR per dwelling, which covers up to one-third of the cost of complete modernization. Like the ANAH system for landlords, it is targeted on improvements in basic 'comfort' – bathroom, toilet, heating.

The 1989 Act is an important piece of legislation, which will influence the direction of housing policy for the rest of the 1990s. It is, of course, easier to pass legislation than to achieve results on the ground. Nevertheless, the Act is important in that it lays down principles and introduces procedures which enjoy broad political support. The Act is not influenced by either side of the doctrinaire 'State versus the market' controversy which has dominated housing policy in some countries. In the French version of *soziale Marktwirtschaft* it is accepted that owner-occupation, private tenancy, and the market are the basic elements in the housing system. At the same time, it is strongly held that public authorities and non-profit housing agencies also have an indispensible role to play, and that public expenditure is needed in cases where the market system fails to work efficiently or equitably. The Act makes an unequivocal commitment to channel more resources to the homeless and badly-housed, which is bound to have an influence on policy.

The detailed provisions of the Act might perhaps be described as mere common sense, but of a type which in the past was sometimes more honoured in the breach than in the observance. The underlying theme is a concern with the end-products of public expenditure, and a readiness to use any channels to achieve the best value for money, whether this means assistance to private landlords, non-profit agencies, targeted assistance with home-ownership, or direct payments to homeless people. There is also a recognition that housing policy, urban renewal policy and social policy cannot be treated as separate activities. Housing policy should seek to avoid 'ghettoization'; urban renewal should not ignore the effects of low-income tenants; to eliminate homelessness, cheap housing has to be available, but the homeless people often need other forms of help as well.

On these issues there is at present a considerable degree of consensus in France. There is widespread criticism of the present Government, and of mainstream political parties in general. The Opposition parties also criticize the Government for not doing enough – for the middle classes rather than for the homeless. There is, however, little serious dispute with the broad lines of the eclectic, pragmatic housing policy which has been pursued for the past fifteen years. Current housing problems require detailed work at the local level rather than dramatic changes in national policy.

## CONCLUSIONS AND PROSPECTS

Housing conditions in general have improved steadily over the past twenty or thirty years, but problems remain. Today, the housing situation in general is quite good, which provides an even greater contrast to the situation of homeless people, or poor people in poor housing. However, homelessness is only the extreme case of a spectrum of diverse housing problems. The 'Law Scrivener', which provides guidelines to lending agencies, with the aim of preventing families incurring too large a burden of debt, illustrates one current type of 'housing problem'. Others are concerned with general conditions in the post-war suburban estates, which include many young people, and are still more or less separate from the 'town'. They lack many of the elements which make up a town, including employment. The way to achieve more social peace is to increase the feeling of 'citizenship' among the population in these districts. Housing is only one of the issues involved, alongside town planning, social policy and economic development; if any one of these elements is lacking, the people in the suburban estates will still feel left out.

A recent report (Delarue 1991) has examined the results of a decade of policies designed to improve housing conditions in the suburbs. It made clear that much has been achieved through the policy of DSQ (*Développement Social des Quartiers*). Between 1984 and 1988, 150 towns and other local authorities undertook programmes to improve 148 estates, including 43,000 dwellings and 1.36 million inhabitants. A total of 9 billion francs was spent in renovating 170,000 dwellings, while 3 billion francs was spent in 6,600 different projects in the social, economic and cultural fields. The official target is to renovate a million HLM dwellings over the period 1989 to 1994. Other programmes include the improvement of education in 'special education sectors'; this programme has been sponsored by a new 'City minister', who is responsible for developing inter-ministerial action at the national and local level.

In spite of considerable achievements in the field of housing, however, it would be wrong to end on too complacent a note. Whenever there are troubles in various suburban estates, the cry goes up to 'improve housing conditions'. There is indeed much that could still be done to improve the habitability of these estates and to make a reality of the 'right to housing' guaranteed by law. A high level of construction and an active renovation programme must be continued, and the legal provisions concerning homeless people must be implemented.

Underlying 'housing problems', however, are more intractable social and economic problems. The rise in the number of homeless people reflects, among other things, a process of industrial restructuring, accompanied by job losses. The persons affected include unskilled young people and 'redundant' older people, who have lost the protection of a family home. France today faces the problems of a persistently high level of unemployment; a hard core of poverty in an affluent society (in spite of the establishment of RMI, the minimum income); racial tension, in a country which used to pride itself on its racial harmony; and a feeling that society has become more violent. (Whether society has actually become more violent is difficult to establish empirically.) France is not alone, however, in facing these problems.

## REFERENCES

Assemblé Nationale (1991) *Perspectives du secteur du logement et de son financement 1992–1995*, October.
Curci, J. (1989) 'Housing Survey 1984', *Données Sociales*, INSEE.
Curci, J. and Taffin, A. (1991) *Enquête Logement*, 'Housing Survey 1988', *Economie et Statistique* no. 240, February, INSEE.
Delarue, J.M., (1991) *Banlieues en difficulté; a relégation*, Paris: Editions SYROS.
Ministére du Finance (1975) *Commission d'Etude d'une Reforme du Financement du Logement* (Barre Report), Paris: *Documentation Française*.
Ministére du Finance et Ministére de l'Equipement (1975) *L'Amelioration de l'Habitat Ancien* (Nora Report), Paris: Documentation Francaise.
Pearsall, J. (1984) 'France' in Wynn, M. (ed.) *Housing in Europe*, London: Croom Helm.
*Rapport Brecinszki* (1986) Paris: Documentation Francaise.
*Rapport Bloch Laine* (1989) Paris: Documentation Francaise.
*Rapport Lebégue* (1991) Paris: Documentation Francaise.
*Rapport Petrequin: Bilan et les perspectives d'évolution du Logement en France* (1989) Conseil Economique et Social.
Willmott, P. and Murie, A. (1988) *Polarisation and Social Housing*, London: Centre for Policy Studies.

# 7 Low income and housing in the Dutch welfare state

*Jan van Weesep and Ronald van Kempen*

## INTRODUCTION

During the past forty years, the government of the Netherlands has adopted a range of programmes aimed at providing all households with a decent income, irrespective of their position in the labour market. Working people enjoy the benefit of minimum-wage laws and protection against the financial effects of sickness, disability and unemployment. Those who cannot work, or no longer work, receive welfare, social security, or help through a host of other programmes.

However, not everyone is equally well taken care of in the Dutch welfare state. While the range of incomes may not be so large as elsewhere in Western Europe, many people have a relatively low income. In fact, Dutch society now shows signs of social polarization. Those dependent on transfer payments, as well as the workers in minimum-wage jobs, have not been able to keep up with the general increase of affluence. While the 4.9 million strong working population earned on average 38,700 guilders in 1986 (up 5.2 per cent from 1984), the almost three million non-active received on average 20,400 guilders (down 10.5 per cent from 1984) (Nota Inkomensbeleid 1989).

Nevertheless, the promotion of income equality remains a high priority in Dutch government policy. In numerous fields, the government also intervenes directly to guarantee equal access to goods and services e.g., in education, health care, and social services. Housing is considered a critical element in the well-being of people and, consequently, another major component of the Dutch welfare state has been the expansion of the social housing sector. In addition, an elaborate system of housing market regulations covering all housing sectors was introduced in the post-war period to safeguard an equitable distribution of housing. A severe housing shortage had to be dealt with, and it was felt that without deep government involvement this crisis could not be resolved.

Over the period from 1945 to 1985 this policy led to the creation of a comprehensively regulated and subsidized housing system that is unique in Western Europe. Yet the housing situation of low-income households is in many respects still worse than that of higher income groups, and new dangers are emerging as the government sets out to deregulate the housing market, and to disentangle itself from what has become a budgetary millstone.

This chapter discusses the relationship of household income, housing situation, and housing market position in the Netherlands. (The distinction between 'situation' and 'market position' is that between a household's actual housing and the housing which – given its income – it would be in a position to acquire. The assumption, often valid in the Netherlands at the present time, is that the housing market is not in 'equilibrium'.) The relationship is set in a context of social, economic, and political structures and grows out of the national housing policy of the post-war period. Thus the historical background sets the stage for current housing issues. The following discussion focuses on the housing situation of low-income households in the four largest Dutch cities: Amsterdam, Rotterdam, The Hague, and Utrecht. This is where low-income groups have become a significant population segment, and where the housing crisis continues unabated. The position of low-income households has long determined the nature and purpose of local housing policies in Dutch cities. In this discussion, we are especially eager to determine whether or not there is a (potential) conflict between high- and low-income groups in the urban housing market. Are the low-income groups being squeezed from their niches?

To provide the proper perspective, the main characteristics of Dutch society are first outlined; there is then a discussion of national housing policy as it evolved during the post-war period and the changes it is currently undergoing. When the discussion shifts to the local level, several concrete questions are posed. What kind of housing is available to low-income groups, and what are their current housing situations? Why and how do different socio-economic groups compete for the same housing? How can their housing market behaviour be explained? How is it constrained?

# MAIN CHARACTERISTICS OF DUTCH SOCIETY

## Demographic backgrounds

After World War II, the population of the Netherlands increased rapidly, because of the high birth rate and the relatively low death rate. This population growth necessitated a rapid expansion of the housing stock, as it compounded the acute housing shortage brought about by war damage and the standstill of construction during the war years. The housing shortage remained a major problem during the 1960s and early 1970s because of the shifts in housing demands that accompanied the increase of personal incomes during the economic expansion of the period. Eventually, the quantitative housing shortage was resolved, only to make room for a growing mismatch of available housing and the housing needs of the population – a qualitative shortage (Van Weesep 1984).

The rate of population increase slowed considerably after the mid-1960s in what has become known as the (second) 'demographic transition' (Van de Kaa 1987). This term applies to the demographic consequences of important social and cultural changes that induce people to move away from parenthood and marriage. Secularization and the seeking of self-fulfilment – the desire to realize more of one's own potential – underlie the break with traditional values. Individuals of both sexes now strive to earn a personal income, which has caused a move away from the traditional family. These social and cultural changes may not be entirely independent of economic trends, but seem to be remarkably insensitive to economic recessions and crises (Van de Kaa 1987).

Longevity, affluence, female emancipation, and a number of other socio-cultural developments have led to a precipitous drop in birth rates, a relative decline in the number of families with children, and a concomitant increase in the number of single persons and two-person households and the number of elderly people (Table 7.1). The rapid growth in the number of households indicates a rapid increase in housing need. In addition to the growth in household numbers, there has been an increase in the variety of living arrangements within specific age groups and types of households. Singles may be of all ages, two-person households may be composed of adults only, or an adult plus a child. These demographic and socio-cultural trends are changing the demand for housing.

Another dimension of the increasingly pluralistic Dutch society is the growth of the 'ethnic' population. The colonial legacy, as well as

*Table 7.1* The changing household composition 1971–85
(thousands)

|  | 1971 | % | 1981 | % | 1985 | % |
|---|---|---|---|---|---|---|
| Single persons | 668 | (16) | 1,128 | (22) | 1,531 | (27) |
| Non-family households | 101 | (3) | 287 | (6) | 295 | (5) |
| Married couples | 926 | (23) | 1,154 | (23) | 1,203 | (22) |
| Married couples with children | 2,100 | (52) | 2,216 | (43) | 2,151 | (39) |
| Single-parent families | 219 | (5) | 301 | (6) | 376 | (7) |
| Extended families | 47 | (1) | 15 | – | 10 | – |
| TOTAL | 4,061 | (100) | 5,101 | (100) | 5,566 | (100) |

*Source:* WBO data, Kersloot & Dieleman 1988: 13

international labour migration, have made the Netherlands' population increasingly diverse. During the 1950s, many Dutch citizens of mixed ethnic stock repatriated from the East Indies (Indonesia), accompanied by some 20,000 ethnic Moluccans. In the mid-1970s, the independence of Surinam brought again a large inflow of other ethnic groups. They joined the contract ('guest') workers who were recruited in the 1960s, and who have meanwhile become established as immigrants. Because of subsequent family reunions, the Turkish population had grown to 170,000 by 1988, while the Moroccan population stood at 130,000. In 1988, only 4 per cent of the 14.7 million permanent residents in the Netherlands was of foreign nationality, yet their concentration in the large cities creates a strong visibility. In many of the pre-war neighbourhoods, as well as in some of the more recent ones, the ethnic minorities now account for more than twenty per cent of the population (Van Kempen 1991). Their numbers and their housing market position have become significant factors in housing policy (Van Hoorn and Van Ginkel 1986).

**Socio-economic developments**

The globalization of the economy, entailing the transfer of manufacturing jobs to newly industrializing nations, underlies the increasing social polarization in the big cities. Between 1979 and 1986, manufacturing in the Netherlands shed 100,000 jobs. The number of traditional blue-collar jobs, certainly those that used to provide entry-level positions for an unskilled labour force, declined. In the remaining industrial activities there was a movement away from the work floor to the front office; the proportion of highly paid jobs in manufacturing grew substantially (Elfring and Kloosterman 1989).

During the 1979–86 period, the service sector expanded by over 500,000 jobs. Yet this provided little solace to the redundant industrial workforce; throughout the western world, the same categories of workers are being by-passed by the growth of the service sector. The situation is especially grim for people over forty and for ethnic minorities (Karim 1988). For those with the right skills, the perspectives are much brighter; for them, there are increasing numbers of jobs with good pay (Van Kempen and Van Weesep 1989). On the whole, however, the new service-sector employment provides neither high incomes nor secure career prospects. The unemployment rate has hardly decreased since the recession of the early 1980s, and the proliferation of the low-paid service jobs adds to the number of low-income people (Dijst and Van Kempen 1991).

The restructuring of the economy coincided with the emergence of a crisis in public spending. In the 1970s, the national economy took a turn for the worse, triggered by the oil crises of the decade, and the ability of the national government to support the increasingly extensive welfare programmes dwindled. Growing public finance deficits led to the need for budgetary restraint, and the housing programmes, which had long been spared from cuts, were eventually trimmed also.

At first, economies were sought by means of greater efficiency in the housing programmes. Various general support programmes (notably construction subsidies) were restricted or abolished, including those aimed at boosting the supply of social housing. The budget constraints led to close-targeting the programmes on the people who actually needed assistance, rather than maintaining housing opportunities for all. Eventually, however, the government decided to withdraw further from housing finance. After the mid-1980s, it aimed explicitly at mobilizing private investment for housing for the middle class, while relegating low-income households to existing housing. The belief was that more housing would be made available by increased 'filtering' (i.e. the handing down of housing vacated by higher income groups). This heralded the end of the massive social housing programmes that had become the hallmark of the Dutch housing system up to 1975. This shift was in line with the general political changes of the 1980s.

**The political factor**

Proportional representation and political pluralism rooted in previous denominational strife has resulted in a tradition of coalition governments. In addition, the Dutch political model is essentially corporatist; advisory boards composed of representatives of all sorts of social organizations strongly influence policy making. Consequently, while the various political parties have different philosophies concerning the goals and extent of housing policy, there is a fair degree of continuity in housing programmes, even when government coalitions change.

During the 1950s, the country was governed by the left–centre coalition which created the main institutions of the welfare state. During this period, the social housing stock was expanded greatly, from approximately 140,000 dwellings in 1945 (10 per cent of the stock) to more than 540,000 (25 per cent) by 1960. In addition, an extensive system of price controls, tenure security, and housing allocation was put into place to guarantee the equitable distribution of housing. Social rental housing accounted for 50 to 65 per cent of the annual housing production, while housing subsidies were introduced to stimulate the private sector. Very few units were built without subsidies (less than 5 per cent). Housing subsidies increased from 0.8 per cent to 3.8 per cent of total government expenditure, while all housing expenditures (including construction loans for social rental housing) increased to 10 per cent of the budget.

During the 1960s, when the left–centre coalition had given way to a series of centrist and right–centre coalitions, deregulation became the explicit goal of policy. Non-subsidized housing became more prevalent, but still did not account for more than one-third of new construction. By the end of the 1960s, housing controls were also being abolished in those parts of the country where the housing shortage seemed to have been eradicated (Van Weesep 1984). This trend of deregulation has continued. Decentralization of powers from central government agencies to local government has recently been

*Table 7.2* Housing expenditure 1970–90
(millions of guilders)

|  | Management expenses | Construction subsidies & loans | Rent subsidies | Urban renewal | Total |
|---|---|---|---|---|---|
| 1970 | 155 | 2304 | 10 | 121 | 2590 |
| 1975 | 290 | 4528 | 314 | 358 | 5490 |
| 1980 | 803 | 6280 | 906 | 705 | 8694 |
| 1985 | 1112 | 10150 | 1645 | 1453 | 14360 |
| 1990 | 352 | 8714 | 1790 | 1194 | 12050 |
| 1995* | 325 | 9130 | 2311 | 1353 | 13119 |

* *Projected*
*Source:* Volkshuisvesting 1990

added to the array of measures to disentangle the government from its housing responsibilities. Nevertheless, the size of the social rental sector has continued to expand since 1960, and by 1986 it accounted for 42 per cent of the total stock. While brick-and-mortar subsidies were severely curtailed, the money spent on rent subsidies sky-rocketed. Only recently has the government succeeded in decreasing its housing expenditures in real terms (Table 7.2), cutting the total to 7.7 per cent of its budget in 1990.

## THE STATE, THE HOUSING MARKET AND LOW INCOMES

### The struggle to overcome the housing shortage

Until recently, the struggle to overcome the post-war housing shortage set the housing policy agenda in the Netherlands. World War II had left many homes destroyed or severely damaged, and the stand-still of new construction had aggravated the problem. The 1947 census showed a shortage of some 300,000 dwellings. Moreover, the war had also left a crippled economy, and the government made economic reconstruction its first priority. To this end, stringent controls were imposed on investment, prices, and wages. Low rents, to avert wage demands, were essential for the success of the economic policy. Thus, the rent control imposed at the outbreak of the war was prolonged. To direct most of the scarce investment capital for productive use, the housing construction programme remained relatively modest throughout the 1950s (Nota Volkshuisvesting 1972).

The post-war housing policies ushered in a new era. Before the war, the construction of social housing authorized by the 1901 National Housing Act was used only as a stopgap. Private housing investment had been the mainstay of housing policy. However, now that rent control reduced the return on housing investment, social housing achieved a new stature. In 1949, 31,000 social rental dwellings were constructed by local governments as compared with 10,000 privately funded dwellings. This dominance of the social sector diminished thereafter (Table 7.3), but only because the government resorted to subsidizing private housing construction. Non-subsidized housing did not account for even a quarter of all housing starts until 1961, and the 35.2 per cent reached in 1963 remained the high-water mark until the late 1980s.

The policy proved to be successful. For each year after 1952, the target of 55,000 new dwellings was surpassed. None the less, the

*Table 7.3* Completed dwellings by financing category 1941–80

|          | Total Completions | Social  | Subsidized | Non-subsidized |
|----------|-------------------|---------|------------|----------------|
| 1941–50  | 186,445           | 107,247 | 79,198     |                |
| 1951–60  | 715,959           | 408,140 | 273,192    | 34,627         |
| 1961–70  | 1,068,839         | 465,231 | 332,161    | 271,447        |
| 1971–80  | 1,236,790         | 417,413 | 540,414    | 278,963        |
| 1981–90* | 1,101,000         | 447,900 | 422,500    | 230,600        |

* included 1990 figures still preliminary
*Source:* Statistical Yearbooks; Volkshuisvesting 1990

housing shortage did not diminish; in 1960 it proved to be exactly as large as it had been in 1947. The changing demographic structure led to a substantial drop in the average number of persons per dwelling (from 4.33 in 1947 to 3.88 in 1960, and 2.56 in 1990). As a result, the housing shortage became an increasingly compelling political issue, leading to heated debates on the annual housing production quotas and occasionally to the sacking of a government.

After 1960, the housing policy envisaged the return to a free market, but this long-term goal was considered attainable only after the successful eradication of the housing shortage. Thus, instead of abandoning housing controls, production incentives were increased. The construction of new housing soared: it reached 100,000 completed units in 1964 and topped 150,000 units in 1972. Some relaxation of rent control and the introduction of new subsidy programmes to promote the owner-occupier sector made investments in housing flow freely. When the market for owner-occupier dwellings weakened in 1968 (and collapsed in 1979), the government reacted by increasing appropriations and permits for social rental housing and subsidized private housing. The building boom thus continued, producing many of the monotonous residential areas with a high proportion of high-rise, multi-family structures. Now, barely twenty years later, these housing estates show severe social and management problems, just like much older housing complexes.

By 1975, it was officially assumed that the numerical housing shortage had been overcome, as the vacancy rate had risen to over 2 per cent. The frantic building of the early 1970s had apparently accommodated the 100,000 households that were looking for a dwelling at census time in 1971. However, there has been resounding talk of a 'new housing shortage' due to the explosion of demand from those types of households whose number was boosted by the demographic transition, and who were not included in the housing demand equations. Young singles demanded housing rights, and became qualified for independent housing; the number of foreign immigrants was grossly under-estimated; and changes in social programmes contribute to a greater demand for independent housing by the elderly than anticipated. Along with the exploding number of divorcees, all these groups operate predominantly in the tightening social rental market. The demand of these households with their weak housing market position is most vigorous in the largest cities. It is here that the effects of demographic and socio-economic changes are the hardest to resolve, given the policy perspectives discussed below.

### Housing allocation and the misappropriation of the social rental stock

An essential ingredient of the post-war housing policy was the controlled allocation of new homes and vacant dwellings to qualifying households. Most of the rental housing was directly allocated by the housing authorities to households that met certain criteria. On the one hand, this system was meant to protect local people from competition from outsiders. At least one member of a household had to have local economic or social ties to be able to obtain a 'residence permit'. On the other hand, allocation was meant to ensure the equitable use of available housing, bolstering the chances of people with a weak housing market position. This also promoted the efficiency of the system, ensuring the optimal use of housing and subsidies. Through the active allocation of vacant housing, municipal housing authorities tried to match the housing needs of home seekers with the characteristics of the available dwellings. The household composition and the size of the dwelling were balanced, as were income and the price of housing. In principle, inexpensive dwellings were reserved for low-income households.

The legal foundation for this system was given in the Housing Allocation Act of 1947, under which each household was required to obtain a permit to use any dwelling – even (officially) if the household owned it. Once access was gained, however, there was no subsequent testing to determine whether the dwelling remained appropriate for the household. This statutory tenure protection system is responsible for the misappropriation of inexpensive rental housing (and subsidies!) by households with an above-average income. Their failure to move on to more expensive homes limits the housing options for low-income households. A recent proposal to introduce income testing for sitting tenants provoked a political fire storm, and such a measure is still not politically acceptable.

Allocation controls were not equally strict for all types of housing. Access to the social rental sector and to other rental units to which the government contributes an operating subsidy was more strictly controlled than access to the owner-occupier sector. But even the use of private rented housing was governed by the regulations. These stipulated exactly to what categories of households a landlord was allowed to rent the dwelling, or which criteria an owner needed to meet to be able to reclaim his property for his own use (Nauta 1979).

By the end of the 1960s, when the housing shortage had diminished in large parts of the country, deregulation set in. The 1947 Act was de-activated in an increasing number of small- and medium-sized municipalities, mostly in the periphery of the country. This deregulation process was abruptly halted in 1974 when a new left–centre government sought to retain control over the housing stock, predominantly to protect the position of low-income households. All expensive dwellings, both rental and owner-occupied, were thenceforth decontrolled, but the municipalities retained authority to prescribe the use of low-cost housing. In recent years, the protected segments of the market have been increasingly narrowly defined, so as to focus housing assistance on those who really need help in finding affordable, decent housing. All others are relegated to the market. For them, the housing authorities are no more than a clearing house, promoting the flow of information about vacant dwellings and homeseekers.

The deregulation runs counter to the need to cut the subsidy budgets. Low-income and middle-income households can obtain more expensive housing, and are then entitled to rent subsidies. It is ironic that a government looking for increased privatization now finds itself in a dilemma; for budget reasons, it feels itself compelled to issue stricter directives on housing allocation. Recently, the allocation rules have become much more stringent with the objective of reducing the need for individual rent subsidies. Low-income households are now virtually barred from new, more expensive rental housing if this would qualify them for a relatively large monthly subsidy payment. This poses a second dilemma for the authorities, since the mixed population of low-cost housing has helped to avoid stigmatization of social housing, and has promoted relatively mixed neighbourhoods (Murie 1990).

The national government itself is also involved in other ways in local housing allocation. It has the right to dispose of up to 10 per cent of new social rental housing units for special groups. Formerly, these dwellings were reserved for national government employees or for people working for (semi-)government agencies. Since 1951, one half of all units claimed under this programme have been set aside for repatriates from the former colonies, and in recent years, for political refugees (Gerrichhauzen 1980). This provision provides the flexibility to transfer key personnel to other locations. It is also a unique instrument to implement an affirmative action policy to house minorities, and to promote their integration within Dutch society.

Housing allocation rules and statutory tenure protection thus modify the relationship between income and housing situation, but power, financial resources, and opportunities are unequally distributed, even in the Dutch welfare state, and income remains a main factor in housing choice (Dieleman 1986). However, the 'Iron Law' of the housing market ('the lowest income households live in the poorest housing, and high-income groups occupy the best') does not apply to its fullest extent. This situation serves the equity principle, but at the same time it may imply direct competition between the affluent and the poor for the same housing, which may not be in the best interests of those low-income households who still need to conquer their place in the housing system.

### Subsidies and the cost of social housing

Since the 1950s, 'brick and mortar' subsidies have played an important role in the Dutch housing system. Because rents were frozen, the necessary construction effort to deal with the housing shortage could be made possible only by subsidizing investment costs. The government provided the capital for the construction of social rental housing in the form of subsidized loans to local authorities and non-profit housing associations. Operating subsidies were also provided to these organizations, and to private investors in rental housing.

The basis for the subsidy was the 'cost rent', derived from construction costs and the current interest rate of a standard annuity loan. However, the rent actually charged was based on complex norms defined by the government, reflecting consumption standards. The difference between the cost rent and this contract rent determined the amount of subsidy. Periodic rent increases slowly decreased the amount of the subsidy. Depending on the level of construction costs, it would take 15 to 20 years before the contract rent would equal the cost rent. After that, the owner could make a profit (above the standard rate of return allowed for in the cost rent) as the annual rent increases would make the contract rent higher than the cost rent. The consequence of this system was that the government was paying a subsidy on housing that would provide a profit to the investor for the better part of its economic life. It made rental housing attractive financially, and did result in vigorous building by private investors. But it was not financially sound for the government, and therefore the system was changed in the mid-1970s.

In 1975, the infamous system of 'dynamic cost rents' was introduced. Its principle is fairly simple. It took future profits (during a fifty year period!) into account when determining the initial subsidy, which therefore fell. The operating losses in the early years could be covered by a ballooning construction loan in the case of social housing, or had to be assumed by the investor in the private sector. Eventually the investor would be compensated by later operating profits. However, the system depended on the exact prediction of the operating costs for the entire fifty year period. This entailed the precise forecast of the rate of inflation and of operating expenses, and it also committed (future) governments to pay subsidies on a given dwelling. This involved too much uncertainty for private investors, and in combination with the lower initial yields, this caused many to withdraw from the housing market. The problems for the government were compounded by the fact that the economy did not perform as expected after 1980; inflation fell, leading to a much slower rate of rent increases, and consequently much higher subsidies. The system was therefore abolished without much ceremony, and replaced by operating subsidies based on agreed budgets. At the same time, the construction subsidies were mostly replaced by personal housing allowances, but future housing budgets will show the past commitments for many years to come. Other construction subsidies, notably for the promotion of owner-occupier dwellings, are also being phased out. And even the subsidy budgets for renovations, site preparation, and development costs are shrinking rapidly.

The decrease in the level of 'bricks-and-mortar' subsidies after the late 1960s implied increasing rent levels for new housing. To keep rental housing affordable for low-income households, rent subsidies were introduced. The programme was initially seen as a supplementary subsidy for the lowest income groups, living in subsidized rental housing built since 1960. Its principle was that the tenant would not pay more than one-sixth or one-seventh part of taxable income for rent.

In 1975, the programme lost its experimental character, when the housing finance system was fundamentally changed to become a mixed system of production and consumption incentives. The contract rents were determined on the basis of what would be affordable for a household with a modal income, and to keep new social rental housing accessible to lower income households, they would be entitled to a personal housing allowance. Eventually this entitlement was extended to all households with a below modal income in low- and moderately-priced housing. Since then, it has become the major form of housing subsidy, and a useful tool to target subsidies to the households that need them most.

The programme was a huge success when measured in terms of the number of households enrolled (Table 7.4), which expanded rapidly, partly because of increasing housing costs, partly because of the stagnation in the growth of personal incomes. Consequently, each year the average housing allowance payment increased, and the budgetary effects became far larger than the government had foreseen. This eventually led to many restrictive changes in the programme, from the introduction of separate payment tables for single persons to the lowering of the rent ceilings of qualifying housing. To reduce the negative effects of these frequent changes on the confidence of tenants and investors, the programme was enshrined in law in 1986.

One of the most recent changes was to cap housing allowance payments. By limiting the monthly payment for newly allocated housing to Dfl. 250.00 (with some exceptions) the government virtually barred low-income households from the newest and the best quality housing, even in the social rental sector. In combination with inducements for higher income households to move to more expensive housing, this may have the unintended effect of increased segregation of socio-economic groups. Therefore the changes were vehemently opposed. A much less controversial change has been that the variations in housing quality were introduced as weights into the subsidy amounts (tenants in higher quality housing received less). Yet even this rule may boost the segregation tendencies.

The individual rent subsidy programme clearly benefited the lowest

*Table 7.4* Individual rent subsidy 1975–89

| Period | Number of recipients | Average subsidy (Dfl) | Total budget (million Dfl) |
|--------|---------------------|----------------------|---------------------------|
| 1975–6 | 348,320 | 974 | 339.3 |
| 1979–80 | 417,903 | 1277 | 533.7 |
| 1984–5 | 715,323 | 1777 | 1271.0 |
| 1988–9 | 907,000 | 1765 | 1600.9 |

*Source:* Papa 1991

*Table 7.5* Monthly housing costs in the rental sector, by income level 1989

|  | Minimum income | | Modal income | | 2–3 × modal income | |
| --- | --- | --- | --- | --- | --- | --- |
|  | Dfl | % | Dfl | % | Dfl | % |
| Net rent | 301 | 20.2 | 435 | 16.8 | 567 | 14.2 |
| Utilities | 158 | 10.6 | 174 | 6.7 | 167 | 4.2 |
| Taxes/rates | 37 | 2.5 | 38 | 1.5 | 37 | 0.9 |
| Total net income | 1493 | 100.0 | 2584 | 100.5 | 3981 | 100.0 |
| Monthly subsidy | 75 | | 7 | | 0 | |
| Share of tenants receiving subsidy | 51% | | 7% | | 0% | |

*Source:* Volkshuisvesting 1990

income groups. In 1989, just over half of all the households with a minimum income were enrolled in the programme, receiving an average payment of Dfl. 75.00 per month. However, it has not rectified the major financial inequity of the housing system, namely, that the lower income groups pay a disproportionate share of their income for housing (Table 7.5). Minimum income households still pay almost twice as much for housing (in relative terms) as households earning two or three times the modal income.

When calculated in a different way, the inequities of the rent costs also show up (Table 7.6). Only 17 per cent of the minimum income households in rental housing have a rent ratio (share of taxable income devoted to rent) of less than 15 per cent, while 77 per cent of households earning two or three times the modal income have a rent ratio of this low level. Conversely, 51 per cent of the minimum income households in rental housing spend over 20 per cent of their income on rent, against 3 per cent of the higher income group. This is partly the result of the presence of many households with higher incomes in low-cost rental housing. Too few tenants move out when their incomes increase. They are either satisfied with the housing they occupy, or there are bottlenecks in the market that preclude adequate filtering. This bias in housing tenure has now become the major issue in housing policy. It leads to the misappropriation of subsidies on the one hand, and to blocked access to inexpensive housing by low-income households on the other. The adaptations of the rent-subsidy programme have not resolved this issue.

The changes in the programme have had another negative effect. They have virtually abolished the right to housing for young, single adults, which they received in the mid-1970s. Their entitlements have been cut while, simultaneously, their incomes have decreased

*Table 7.6* Rent burden by income level: percentage of tenants 1989

| Rent ratio | Minimum income % | Modal income % | 2–3 × modal income % |
|---|---|---|---|
| 0–10 % | 6 | 10 | 34 |
| 10–15 % | 11 | 29 | 43 |
| 15–20 % | 34 | 31 | 20 |
| 20–24 % | 30 | 22 | 2 |
| 25 % and more | 21 | 9 | 1 |
| Total | 100 | 100 | 100 |
| Average rent ratio | 20 | 17 | 12 |

*Source:* Volkshuisvesting 1990

through changes in the minimum wage laws for young people. The changes were introduced as improvements of the programme but the effect may be that young people will be more reluctant to look for housing for which rent subsidies are needed to keep it affordable (Boelhouwer 1989). That would neutralize some of the most positive effects of the programme in the context of the welfare state: namely, that the housing situation of low-income households is not primarily determined by their income position, but by their housing needs.

**The changing role of housing associations**

The National Housing Act of 1901 enabled local governments to take an active role in the provision of housing. They were allowed to subsidize the construction of 'working class' homes, and would be reimbursed by the national government. While this provision did not immediately lead to the production of many social dwellings, it did lead to a heated debate. Direct public participation in housing construction and management was highly suspect in the eyes of the politicians, since it smacked of state socialism (Van der Schaar 1980). Instead, the non-profit housing associations, the successors of the philanthropic corporations and the mutual (workers') building societies of the late nineteenth century, became the prime vehicle of the social housing programme. Accredited by the government as non-profit organizations to build and manage social (rental) housing, they qualified for low interest construction loans. Any social group or organization was free to organize such a housing association to promote housing for its own constituency. Also municipal housing offices could be set up under the same provisions.

Before the Second World War, the number of associations grew

*Table 7.7* Social housing associations by size of operation 1989 (number of dwellings)

|      | 0–599 | 600–1,799 | 1,800–3,999 | 4,000–9,999 | 10,000+ | Total |
|------|-------|-----------|-------------|-------------|---------|-------|
| 1958 | 870   | 88        | 47          |             | 27      | 1,032 |
| 1967 | 680   | 171       | 95          |             | 63      | 1,009 |
| 1977 | 421   | 249       | 160         |             | 104     | 934   |
| 1985 | 234   | 303       | 213         | 93          | 11      | 854   |
| 1988 | 202   | 284       | 213         | 107         | 14      | 820   |

*Source:* Volkshuisvesting 1990

rapidly, but most remained very small, part-time organizations, with limited numbers of dwellings. This changed after the war, when the social housing programme was beefed up. The introduction of operating subsidies made a big difference, since the lack of these had limited the demand for social housing before. As we showed earlier (Table 7.3), the construction of social housing units increased rapidly, as did their share in the total construction programme. The housing associations grew larger and their numbers declined in response to the need to achieve economies of scale and to maintain a professionally trained organization to deal with the responsibilities thrust upon them (Table 7.7).

Initially, the private associations and the municipal housing offices worked on an equal footing. Until 1960, the number of dwellings constructed by each was almost equal. During the 1950s, the municipalities constructed a total of 208,000 units; the private associations built 187,000. The big cities were especially active. Then the first effects of increased privatization were recorded. During the 1960s,

*Table 7.8* Relative positions of housing associations and public housing offices 1988

|                                        | Housing associations | Public housing offices |
|----------------------------------------|----------------------|------------------------|
| Number of dwellings ($\times$ 1000)    | 1,650                | 300                    |
| Book value of property                 |                      |                        |
|   (total $\times$ billion Dfl.) | 83                | 11                     |
|   (per unit, $\times$ 1000 Dfl.) | 50               | 36                     |
| Financial reserves                     |                      |                        |
|   ($\times$ billion Dfl.)    | 7.6                  | 1.1                    |
|   (% of annual rent collection) | 82                | 74                     |
|   (% of book value)          | 9                    | 10                     |
| Operating profits (billion Dfl.)       | 0.8                  | 0.14                   |

*Source:* Nota volkshuisvesting 1989: 58

the housing associations built 50,000 dwellings more than the munici-palities. At the end of the decade, the privatization was underlined when a government directive restricted the housing efforts of local governments to those situations where private non-profit associations were unable to carry the responsibility for the production of social housing. The effects showed clearly in the construction figures for the 1970s: the non-profit associations built over 430,000 units during the decade, the municipal housing offices only 73,000. The 'public' housing stock slipped in quality as it was no longer rejuvenated. In addition, the municipalities acquired large numbers of derelict pri-vate dwellings as a consequence of their urban renewal programmes. Many of these dwellings were eventually renovated, many others demolished. Consequently, a major difference emerged between the 'public' and the 'private' social rental stock (Table 7.8).

Table 7.8 indicates that the financial position of the housing associations is sound enough to meet the challenges of further privatization. Their financial reserves are substantial, and so far they have made operating profits. In addition, they should be able to generate the investment funds needed to keep up the quality of the social rental sector through loans in the capital market, since their obligations remain fully guaranteed by the government. The newly instituted mutual fund to aid individual associations with financial problems provides additional security for their continued solvency. The great majority of associations seem well prepared to face the increased financial independence which the privatization programme of the government has in store for them (Nota Volkshuisvesting 1989). This means that they will continue to provide housing for low-income households. However, it remains to be seen whether these households can fully use these opportunities, given their housing market positions and the competition from other groups.

## THE HOUSING MARKET POSITION OF THE URBAN POOR

### Housing market positions

Even in a welfare state, the key to anyone's position in the housing market is found in his or her socio-economic position. Yet, as we have seen above, household characteristics other than income influence households' preferences and housing demand. The ability to realize housing aspirations is circumscribed by the state of the housing market; moves are often postponed when preferred housing is not available. Statutory rules and regulations, the availability of subsidies, and the characteristics of the housing stock modify housing situations and the housing market position of various population groups (Van Kempen *et al.* 1989).

Analysts have shown that the lack of owner-occupier housing and the relative scarcity of more expensive rental units create bottlenecks that prevent adaptive moves of higher income households. The need to promote filtering to keep the system in better balance suggests that there should be an emphasis on the construction of more expensive dwellings in the cities. An increased share of such housing in the new construction programmes would also be in line with current trends in national housing policy, emphasizing private investment and deregulation. Yet the residential construction programmes of the large cities still show how the local authorities attempt to side with the lowest income groups by building dwellings to meet their needs directly. Whereas social rental housing has typically made up 30 to 45 per cent of new construction nationally since 1980 (somewhat less than during the 1970s), the corresponding figure for the large cities is consistently much higher.

*Table 7.9* New construction by tenure and financing category 1988

| | Netherlands | Amsterdam | Rotterdam | Hague | Utrecht |
|---|---|---|---|---|---|
| *Total completions* (=100%) | *118446* | *5196* | *4882* | *3426* | *1134* |
| | % | % | % | % | % |
| Social (rental) housing | 33.9 | 82.5 | 53.1 | 72.8 | 61.9 |
| Subsidized housing | | | | | |
| rental | 8.7 | 4.7 | 1.7 | 6.2 | 6.5 |
| owner-occupier | 35.7 | 7.1 | 39.6 | 16.7 | 30.7 |
| Non-subsidized housing | | | | | |
| rental | 0.9 | 1.3 | 1.4 | 1.5 | 0 |
| owner-occupier | 20.8 | 4.4 | 4.2 | 2.8 | 0.9 |
| TOTAL | 100.0 | 100.0 | 100.0 | 100.0 | 100.0 |

*Source:* Volkshuisvesting 1990

This priority shows clearly in the differentiation of the 1988 programme (Table 7.9). The emphasis on social rental housing is clear from these figures. The situation in Rotterdam is somewhat less extreme than in the other large cities; but even here, the overwhelming majority of new dwellings is subsidized.

Much of the new housing in the cities has been built in urban renewal areas. These reconstruction programmes still demonstrate the principles introduced in the 1970s: new housing has to meet the needs of the (low-income) population living in the neighbourhood before its renewal. The preferences of people with growing incomes, leave alone those of outsiders who might wish to move to the cities, are not sufficiently reflected in the residential construction programme. They are relegated to the more expensive existing housing outside the allocation controls exercised by the authorities; such housing remains scarce in the big cities because of rent control and restrictions on conversions and upgrading. Middle- or higher-income households often have no realistic alternative but to move to a suburban community, or to stay put if they already live in inexpensive housing.

The housing stock is segmented by age, tenure, type and, not least, by price; different segments offer varying possibilities for different types of households. The role of the various segments depends on the number of households that aspire to a certain kind of dwelling, but also on developments in other segments within the same housing market area. When, for example, old and inexpensive dwellings are subjected to urban renewal, low-income households may be compelled to seek inexpensive existing housing in other areas (Hoogvliet & Jobse 1989; Briene *et al.* 1989). This explains in part the recent changes in the population of the early post-war areas, which supplies an increasing share of the low-cost rental housing stock in the large cities (Van Kempen 1991).

When a household moves, its previous dwelling can become available for another household. Therefore, especially where the present housing distribution is inequitable, the promotion of filtering moves up on the political agenda. The alleged mismatch of incomes and housing costs, particularly the widespread use by higher income groups of inexpensive, subsidized rental dwellings, has become a major issue. Such use is considered a misappropriation of scarce housing resources, forcing low-income households to resort to more expensive homes or even postpone household formation. It is therefore an important empirical question to determine whether or not the lowest and the higher income groups compete directly for the same housing.

### Housing situations

Data from the National Housing Needs Survey (See *Acknowledgement*) can be used to describe the housing situations of the various income groups in the large cities (Van Kempen *et al.* 1989). Low-income households were compared with two other income groups

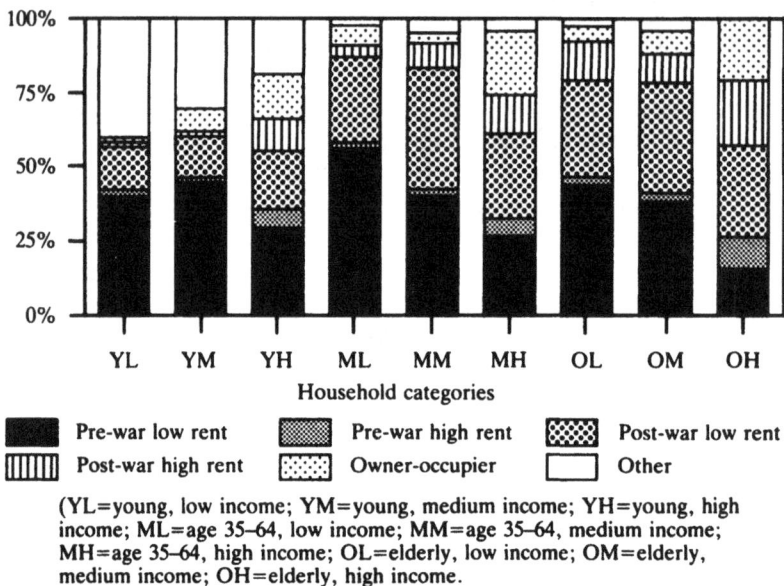

(YL=young, low income; YM=young, medium income; YH=young, high income; ML=age 35–64, low income; MM=age 35–64, medium income; MH=age 35–64, high income; OL=elderly, low income; OM=elderly, medium income; OH=elderly, high income.

*Figure 7.1* Housing situations of single persons by age and income 1985/6

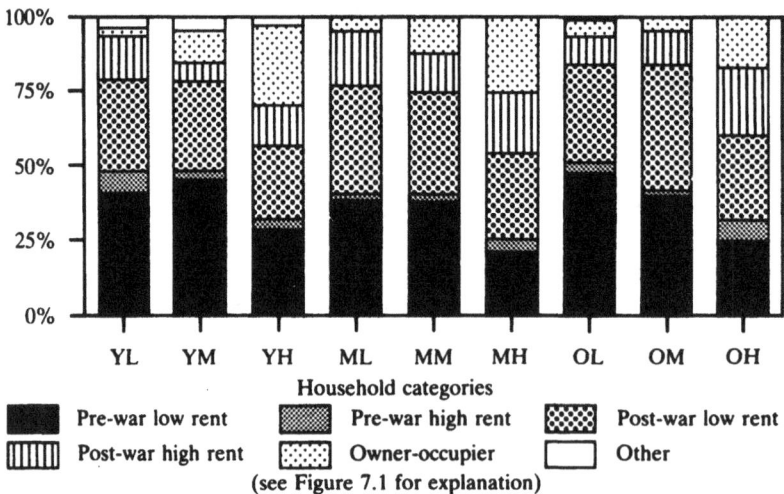

*Figure 7.2* Housing situations of two or more persons by age and income, 1985/6

(households with an income above the minimum but below the median, and households with an above median income). To define the low-income group, the official minimum income level was taken as the ceiling; in 1985, this amounted to Dfl. 13,500 per year for single persons and Dfl. 18,000 for two or more person households. A net annual income of between Dfl. 13,500 and Dfl. 22,000 (Dfl. 18,000 – 30,000 for two or more persons) defined the middle category, while incomes above these levels are defined as the higher income households. Yet, as others have shown, the main differences in housing situations are as much related to age and household size as to income.

Figures 7.1 and 7.2 demonstrate clearly how income influences the current housing situations. Within each age category, differences in income lead to a wide variation in housing situations. Irrespective of other characteristics, households with the lowest income show the lowest proportions of home-owners. Many low-income households live in the inexpensive rental stock in the cities, but differences within this category by age suggest that immobility is an important factor in patterns of housing use. The diagrams also make it clear that the households with a somewhat higher income are very much dependent on the same (broad) housing categories. Even the highest income households are well represented in low-priced housing; in each age and household type, some 50 per cent of the households with an above median income live in either pre-war or post-war inexpensive rental housing. Because of the low rents, their housing costs are very low; this is the misappropriation that plays such a major role in the proposed changes to the present housing system (Nota Volkshuisvesting 1989).

Some other details are also striking, for instance, young single persons are the only category that depends to a large extent on accommodation other than standard dwellings (mobile homes, houseboats, single-room occupancy, loft spaces, etc.). The drop-off with income suggests that this choice is not entirely voluntary, although there is an effect of alternative lifestyles and social arrangements. Low incomes seem to bar the young from standard housing, for some time. The fact that low-income households consisting of two or more persons make much less use of such alternative accommodation can result from their higher priority status in housing allocation, or from age differences.

**Competition in the housing market**

Differences in income are not only responsible for variations in housing situations, but also affect a household's chances of achieving the desired dwelling. Lack of mobility among low-income groups points to bottlenecks in the housing market, which prevent households from finding a more suitable dwelling. The data of the Housing Needs Survey show that, in general, immobility increases with age. The proportion of 'stayers' (households that have lived in the same dwelling for fifteen years or more) is highest among the over-55 year olds. Stayers in younger age groups are rare, but this is clearly a consequence of their rapidly evolving household composition, and their recent entry into the housing market. Nevertheless, the data show that there are apparently opportunities for moves in the urban housing stock. There are also relatively more stayers among the higher income groups than among low-income households. Immobility in this group seems to be a sign of satisfaction (or complacency), rather than dissatisfaction.

The most important question is to determine if higher income households (when they move) move to the same housing segments as lower income households. If they do this would show that the income groups compete actively for the same housing within the urban housing market, either by choice or because of the constraints imposed by the composition of the urban housing stock. This question was examined, with only those who moved at least once during the preceding five years being included in the analysis, and with the elderly and the occupants of non-dwelling types of accommodation left out.

The largest number of moves take place within the rental sector, which constitutes an overwhelming proportion of the housing stock of the large cities. Most low-income households (95 per cent) move exclusively within this sector. Among the households with higher incomes, the likelihood of a move from the rental to the owner-occupier sector is substantially higher. Nevertheless, even among the households with an above median income, 79 per cent moved within the rental sector. This may be an effect of the highly skewed housing stock; in Amsterdam and Rotterdam for example, only 7 and 11 per cent respectively of the stock is owner-occupied, compared with 42 per cent for the Netherlands as a whole. Nevertheless, the fact that higher-income households do not move to the more expensive rental housing is as much a matter of choice as of low mobility rates. Of all the higher income households who moved to a rental unit, 70 per cent moved to a dwelling of no more than Dfl. 450 per month. The respective figure for the poorer households was 86 per cent. Obviously, households from different income groups compete actively for inexpensive rental dwellings.

*Table 7.10* Present and desired type of dwelling; households in the four big cities by income group 1985/6

| Present | Desired | Minimum or below | | Minimum to median | | Above median | |
|---|---|---|---|---|---|---|---|
| | | No. | % | No. | % | No. | % |
| Rent <450 | Rent <450 | 30120 | (85) | 36030 | (85) | 34460 | (55) |
| Rent <450 | Rent >450 | 4290 | (12) | 4320 | (10) | 16400 | (26) |
| Rent <450 | Owner-occ. | 1090 | (3) | 2150 | (5) | 11420 | (19) |
| Total | | 35500 | (100) | 42500 | (100) | 62280 | (100) |
| Rent >450 | Rent <450 | 3160 | (62) | 4790 | (61) | 2390 | (11) |
| Rent >450 | Rent >450 | 1650 | (32) | 2630 | (33) | 9800 | (44) |
| Rent >450 | Owner-occ. | * | * | * | * | 10060 | (45) |
| Total | | 5120 | (100) | 7800 | (100) | 22250 | (100) |
| Owner-occ. | Rent <450 | * | * | 1970 | (50) | 730 | (5) |
| Owner-occ. | Rent >450 | * | * | 760 | (19) | 2052 | (13) |
| Owner-occ. | Owner-occ. | – | – | 1230 | (31) | 13050 | (82) |
| Total | | 670 | (100) | 3960 | (100) | 15832 | (100) |

* suppressed; – none
*Source:* WBO 1985/86. Previous and actual occupants of non-dwelling types of accommodation excluded

Income becomes an important variable, however, with respect to tenure choice and the rent level of the desired dwelling. The difference between the stated aspirations and the observed behaviour shows that there are bottlenecks in the functioning of the urban housing markets. The higher income households have more ambitions than the other two income groups, to move to a more expensive rental home or to become home-owners (Table 7.10). Of the above median income households 26 per cent who want to move from a less expensive rental unit (< Dfl. 450 per month) are inclined to select a more expensive rental unit, and 19 per cent wish to move to the owner-occupier sector. But even if such dwellings were in ample supply, the competition among the groups would remain strong. The remaining 55 per cent of the higher income households desire another inexpensive rental unit. Because the overwhelming majority of the low-income households also want to move to an inexpensive rental home, we maintain our conclusion that there is and will be direct competition among higher and lower income groups in the urban housing market.

## THE HOUSING FUTURE FOR LOW-INCOME HOUSEHOLDS

During the past forty years, as the welfare state has developed in the Netherlands, the housing situation of low-income households has significantly improved. At the beginning of that period, private rental housing made up over 40 per cent of the stock, while social rental housing accounted for approximately 20 per cent. At the moment this situation is completely reversed, with the social rental sector accounting for 42 per cent, while private landlords (individuals and institutional investors) operate less than 14 per cent of the stock. Thirty-five per cent of all housing is operated by non-profit housing associations, and 7 per cent by public housing offices. In the large cities the private rental stock still accounts for between 35 and 50 per cent of stock. Even here, however, public control over the use of housing guarantees that low-income households have access. Rent control and various subsidies, including the widespread use of housing allowances, have kept new housing affordable, in spite of the stubborn housing shortages in various segments of the stock.

The large share of the social sector, and its relatively high quality, have helped to avoid stigmatization; social dwellings do not function as housing of last resort. In fact, a substantial part of low-cost rental housing is used by households with median incomes or above, which has prevented excessive spatial segregation. Yet dark clouds are gathering over the social housing sector, especially in the big cities. The threats emanate from two developments: first, a rapid decline in middle-income, and an increase in low-income and high-income groups; second, policy changes leading to a weakening of the housing market positions of low-income households, and to increased segregation.

The first development results from an increase in the number of both low-paid and highly remunerated jobs. At the same time, many jobs are being lost, and the sector of the population that is dependent on various social programmes has increased dramatically over the past few decades. The income position of this sector is being threatened, as the generous support programmes of the welfare state increasingly fall victim to budget cuts. These processes are most apparent in the big cities, where both the well paid and the poorly paid jobs are concentrated. The big cities have more than their fair share of the poor: in 1985, an estimated 17.5 per cent of households in the large cities had an income of less than 15,000 guilders, and 49.5 per cent less than 23,000 guilders; for a country as a whole, these shares were 11.1 and 37.5 per cent, respectively. The elderly, the unemployed, and other categories of non-active people were on average over-represented in the four largest cities. At the same time, the well paid jobs tend to be concentrated in and around the largest cities. Of course, not everybody who works in the city also wants to live there, but recent research indicates that, among higher income households, the city was more popular as a place to live in the 1980s than in the two previous decades.

The increasing polarization of incomes leads to social conflict, which has become more complex as a result of a multitude of demographic developments. In addition to traditional family-type households, new household types have emerged. They, too, gravitate towards the cities, where they find a housing stock that is suitable for their needs, and the facilities to support the lifestyle they desire. Suburban residential environments, by contrast, do not offer much that attracts the new types of household.

The simultaneous increase in the numbers of the poor and the affluent, however, does not necessarily lead to a dual city in the sense of a socially and spatially segregated city, where rich and poor live in isolation from each other. There are still many households with intermediate incomes (Marcuse 1989), but there are clear indications that gentrification is taking place in some neighbourhoods (Van Weesep & Musterd 1991), while elsewhere concentrations of low-income households, including ethnic minorities, are emerging (Van Kempen 1991). There is active competition among the various population groups for the same housing, and low-income households may lose out unless government intervenes strongly on their behalf. Recent policy developments, however, suggest that the government is increasingly reluctant to shoulder this task.

Recent Dutch housing policy proposals (Nota Volkshuisvesting 1989) devote much attention to the bottlenecks in the housing system, and the ways of correcting alleged inequities. The proposals are suspected by many people of having been instigated more by the need to cut the budget than by the desire to solve the real housing problems of the people. Even though many low-income households live in the least expensive housing, and many benefit from housing allowances, large numbers still have exorbitant housing costs (Van Kempen *et al.* 1989). The major obstacle to a more equitable distribution of housing costs is considered to be the immobility of higher income households living in low-rent social housing. Many of the proposals therefore aim to promote filtering.

The most controversial proposal is to introduce income testing, so that the leases of higher income households can be terminated. Another proposal, which has run into vigorous opposition, is the sale of social housing to sitting tenants. If these proposals are unacceptable, it follows that there are only two ways to make inexpensive housing available to lower income households: to build new low-cost dwellings; or to tempt the higher-income groups to move out. The first is not very likely, as future housing construction programmes foresee a much reduced share for social rental housing. The second may be facilitated by the current emphasis on owner-occupier housing, to be constructed without subsidies. These much desired new dwellings should promote the housing opportunities for the poor by filtering in the existing stock, but there is a danger of increased spatial segregation and the weakening of the housing market position of the lowest income groups.

Given the large and increasing numbers of low-income households that need to be accommodated urgently, the cities have tended to retain their emphasis on the construction of social rental housing. Yet, with rising interest rates, high construction costs, and diminishing subsidies, this may not bring a permanent solution to the plight of the low-income groups; they will depend increasingly on the ever tighter housing allowance programme, and pay high housing costs to boot.

Many of the new proposals intended to make the housing system more 'efficient' (that is, to decrease the burden of subsidies) entail the danger of increased segregation of rich and poor. Housing policy should still seek to avoid such segregation. Even though the sharing of a territory by high- and low-income groups does not necessarily lead to intensive social contact and such social contacts as exist do not always promote cohesion, housing policy should aim at more than the efficient use of ever more limited resources. It should aim to promote the integration of society, not the development of ghettos for the poor. No one wants the emergence of a Dutch 'underclass'.

*Acknowledgement*

This chapter is partly based on the contribution by the authors to the national research programme 'Urban Networks'. The project from which data were used was sponsored by the Ministry of Housing (DGVH), which kindly provided the data from the national Housing Needs Survey (WBO).

**REFERENCES**

Boelhouwer, P.J. (1989) *De individuele huursubsidie en het woningmarktgedrag van huishoudens*, Delft: Delftse Universitaire Pers.
Briene, M.F.M., Dieleman, F.M., Jobse, R.B. & Floor, J. (1989) *Beheer en verhuurbaarheid; een empirische studie van enkele naoorlogse woningcomplexen*, Rotterdam/Utrecht: IOP-Bouw/Instituut voor Ruimtelijk Onderzoek.

Dieleman, F.M. (1986) 'The future of Dutch housing: a review and interpretation of the recent literature', in *Tijdschrift voor Economische en Sociale Geografie 77*, 237–45.

Elfring, T. & Kloosterman, R.C. (1989) *De Nederlandse 'Job Machine'; de snelle expansie van laagbetaald werk in de dienstensector, 1979–1986*, Amsterdam: Economisch Geografisch Instituut Universiteit van Amsterdam.

Gerrichhausen, L.G. (1980) 'Voorkeurs woningen, beleidsevaluatie', in *Beleid en Maatschappij* 10: 282–91.

Hoogvliet, A. & Jobse, R.B. (1989) 'Naar een evenwichtiger opbouw van de bevolking', in *Bouw* 9: 35–7.

Karim, P.D.E. (1988) *Arbeidsmarkt perspectief Amsterdamse werklozen*, Leiden: Stichting Research voor Beleid.

Kersloot, J.M. & Dieleman, F.M. (1988) *Van stagnatie tot herstel; de ontwikkeling van de markt van koopwoningen*, Utrecht: Faculteit Ruimtelijke Wetenschappen.

Marcuse, P. (1989) '"Dual City": a muddy metaphor for a quartered city', in *The International Journal of Urban and Regional Research* 13: 697–708.

Murie, A. (1990) 'Housing market developments in Britain; implications research and policy in the Netherlands', in *Stedebouw en Volkshuisvesting* 71: 11–16.

Nauta, A. (1979) *Inventarisatie toewijzingsregels, deel 1: Methodische inleiding op het onderzoek*, Rapport 6024, Rotterdam: Bouwcentrum.

*Nota Inkomensbeleid* (1989) SDU, 's-Gravenhage.

*Nota Volkshuisvesting* (1972) Tweede Kamer der Staten Generaal-zitting 1971–2, 11784, Nr. 1. Staatsuitgeverij, 's-Gravenhage.

*Nota Volkshuisvesting in de jaren negentig van bouwen naar wonen* (1989) Ministerie van Volkshuisvesting, Ruimtelijke Ordening en Milieubeheer, 's-Gravenhage.

Oyst, M.J. & van Kempen, R. (1991) 'Minority Business and the "Hidden Dimension". The influence of urban contexts on the development of ethnic enterprise', *Tydschrift voor Economische en Sociale Geografie* 92: 128–39.

Papa, O.A. (1991) *Vergelijkende studie naar volkshuisvestingssystemen in Europa*, Delft: Onderzoeksinstituut voor Technische Bestuurskunde.

Van de Kaa, D.J. (1987) Europe's second demographic transition, in *Population Bulletin* 42: 1.

Van der Schaar, J. (1980) 'Volkshuisvesting. Ontwikkelingen in beleid en denkbeelden tot de tweede wereldoorlog', in *Stedebouw en Volkshuisvesting* 61: 61–78.

Van Hoorn, F.J.J.H. & van Ginkel, J.A. (1986) 'Racial leapfrogging in a controlled housing market', in *Tijdschrift voor Economische en Sociale Geografie* 77: 187–96.

Van Kempen, R. (1991) *Lage-inkomensgroepen in de grote stad: woonsituatie, mobiliteit en woonwensen in Amsterdam en Rotterdam*, Stepro rapport No. 103c, Utrecht: Faculteit Ruimtelijke Wetenschappen.

Van Kempen, R., Teule, R.B.J. & van Weesep, J. (1989) 'Low-income households and their housing situations in large Dutch cities', in *The Netherlands Journal of Housing and Environmental Research* 4: 321–35.

Van Kempen, R. & van Weesep, J. (1989) 'High incomes, low incomes and the divided city: social polarization in The Netherlands', paper for the 7th Urban Change and Conflict Conference, Bristol, 17–20 September 1989.

Van Weesep, J. (1984) 'Intervention in the Netherlands: Urban housing policy and market response', in *Urban Affairs Quarterly* 19: 329–53.

Van Weesep, J. & Musterd, S. (1991) *Urban Housing for the Better-Off: Gentrification in Europe*, Utrecht: Stedelijke Netwerken.

*Volkshuisvesting in cijfers 1990* (1990) Zoetermeer: Directoraat-Generaal voor de Volkshuisvesting.

# 8 The new housing shortage

## An international review

*Graham Hallett*

### A CONTINUING IMPROVEMENT IN HOUSING CONDITIONS?

Housing conditions in the USA and Western Europe continued to improve in the 1980s – according to the statistics of numbers of dwellings (in relation to households), and the size and equipment of dwellings. In every country, the number of housing units per 1,000 inhabitants has risen steadily since 1975, with 1987/8 figures ranging from 385 in the Netherlands to 456 in France (Table 8.1) The GDR shared in this improvement, with a 1987/8 figure (420 per 1,000) about the same as that for the USA and well ahead of that for the Netherlands.

These figures, however, need to be treated with caution. A substantial number of dwellings may be second homes or situated in areas undergoing depopulation. In his chapter on France, J-P. Schaefer mentions that some unmodernized older dwellings in declining towns (or empty farmhouses in the Massif Central) are effectively outside the housing market. Similarly, Otto Dienemann cites the

*Table 8.1* Housing units per 1000 inhabitants and average floorspace

|  | 1975 No. | 1980 No. | 1987/8 No. | Floorspace* square metres |
|---|---|---|---|---|
| USA | 387 | 410 | 419 | 149 |
| UK | 364 | 382 | 399 | 88.6** |
| FRG | 383 | 412 | 430 | 99.3 |
| GDR | 370 | 401 | 420 | 62.7 |
| Netherlands | 320 | 343 | 385 | 101.0 |
| France | 339 | 436 | 460 | 105.5 |

* latest available year; average useful floorspace per new dwelling.

*Source:* UN.ECE and national statistics: ** Nationwide Building Society; the definition may not be identical

number of dwellings in the former GDR that are virtually uninhabitable. Thus actual shortages and statistical 'surpluses' can exist in a
country at the same time. Moreover, the number of households has
been rising faster than the total population. It is true that, in all the
countries for which statistics are available, the number of dwellings
exceeds the number of households, but these figures must also be
viewed with caution, both because of the geographical/second home
problem and because a group of people living in a dwelling are
counted as a 'household', irrespective of whether they wish to be one.

The average floor space of new housing in the late 1980s was
around 100 square metres in Western Europe; the figure for the USA
(149 square metres) was substantially higher, and the figure for the
German Democratic Republic (62.7) was substantially lower.

The figures for what the French call 'comfort', and the British
'amenities', have shown large, unambiguous improvements. As
Schaefer points out, the percentage of dwellings with toilet and bath
rose from 56 per cent in 1970 to 90 per cent in 1988. Jay Howenstine
points out that in 1987 the percentage of homes in the USA which
were 'sub-standard' was 7 per cent, compared with 49 per cent in
1940. In England (*sic*), the number of dwellings lacking sanitary
'amenities' had fallen to 3.9 per cent (Maclennan and Niner 1990).
The GDR also showed striking improvements, if not to quite the
same level, accompanied by a 'convergence of living standards'
(Table 5.2). In West Germany, the proportion of households without
a bath fell sharply, even in the lowest income quintile (Fig 4.2).

On the physical condition of housing, the situation varies between
countries. For the Federal German Republic, Ulbrich and Wullkopf
revealingly mention that there are no statistics for physical unfitness,
and that this is not a problem. (Indeed, it could be argued that West
German housing is 'too good'!). The Netherlands also haś a relatively
new housing stock, in generally good condition. The UK has an older
housing stock than the continental European countries, and a
significant percentage is in need of substantial repairs. The dwellings
in England which, in 1986, were either 'unfit' (for human habitation,
as the more eloquent older expression put it) or in 'serious disrepair'
was 11.5 per cent of the total, and there had been little change over
the previous decade (Maclennan and Niner 1990).

On overcrowding, all the available national statistics indicate a
significant fall. In France, the proportion of 'overcrowded' dwellings
fell from 24 per cent in 1970 to 12 per cent in 1988 and, in that year,
far more dwellings were 'very under-occupied' than were even
'slightly overcrowded'. In the USA, the percentages of overcrowding

in 1987 had fallen to between 2 and 7 per cent. These overall improvements do not necessarily exclude a worsening for certain social groups. Ulbrich and Wullkopf give a breakdown by the richest and poorest household quintiles, which shows that there was a general improvement in space standards for the poorest quintile, but not for large families.

Finally, the surveys that have been made of 'housing satisfaction' indicate that most people are satisfied. Schaefer quotes a survey showing that only 9 per cent of French people are dissatisfied. Surveys in Britain showed a figure of 5 per cent in 1986; it had fallen from 11 per cent in 1975, and there had been falls in all tenure categories, although private tenants and council tenants were more dissatisfied than owner-occupiers (Building Societies Association 1986, p. 21).

## THE 'NEW HOUSING SHORTAGE'

In spite of these undoubted improvements in housing conditions, there have been continual complaints in recent years about a 'new housing shortage', relating mainly to affordability. Whereas the post-war housing problem had been one of overall physical shortage, the emerging problem, it was argued, was that many households could not afford prevailing rents or house prices. Peter Malpass concludes that, in the UK, 'there is a serious affordability problem, affecting all tenures' and Jan van Weesep argues that, in the Netherlands, (which in the 1980s pursued a very different housing policy from the UK), the quantitative housing shortage made room for 'a growing mismatch of available housing and the housing needs of the population – a qualitative shortage'.

These views were not universally shared. It was often argued by Government spokesmen, building societies, and some academics, in all the Western countries, that there was no shortage and no crisis. And yet most people who observed the increasing number of people sleeping rough in London or US cities could not help feeling that something was wrong. This 'casual empiricism' was supported by statistical evidence of increased numbers of homeless people; of rising repossession (foreclosure) rates in some countries; and of high and rising expenditure on housing by low-income groups as a proportion of income. Were sections of the population facing growing problems of paying for housing, and, in the extreme case, failing to obtain any housing, even though conditions for most of the population were improving?

The contributors of the national chapters conclude that there were indeed growing problems of housing affordability for significant minorities of the population in the USA, the UK, West Germany, France and the Netherlands – although the intensity and the mix of problems varied from country to country. The chapter on the former German Democratic Republic, by contrast, stresses the grave problems of physical deterioration which underlay an apparently egalitarian system, and looks forward to the new affordability problems – as well as the new opportunities – which will arise with the introduction of a more market-orientated system. What were the causes of the 'new housing shortage' in the Western economies in the 1980s? How far can it be attributed to the housing policies adopted? What can be done to alleviate it?

The 'new housing shortage' can be divided into two main categories: (a) homelessness and (b) excessive housing costs in relation to income. These categories are neither mutually exclusive nor comprehensive. Homelessness is sometimes a straightforward affordability problem; sometimes more than this. There are also (in spite of the favourable national statistics) localized issues of overcrowding and of poor or unconventional accommodation, verging on homelessness; these, however, are usually a reflection of the 'unaffordability' of better housing.

### Homelessness

The contributors of the national chapters conclude that, in all the countries studied, there was a rise in the 1980s in the number of people who were homeless, or near-homeless. They also agree that most of the increased numbers of the homeless consisted of young people, and included, for the first time since the War, significant numbers of women and children. It is very difficult, however, to estimate absolute numbers, partly because of the difficulty of defining 'homelessness'. The number of people who are homeless in the sense of 'roofless' is merely the visible part of an iceberg comprising the 'hidden homeless' living in a range of precarious circumstances. Thus Malpass cites a study which estimated that, in 1989, there were 3,000 people sleeping rough in London, but the same study gives estimates of 30,000+ squatting in unoccupied buildings or in short-life housing, 15–17,000 in hostels or 'bed and breakfast' hotels, and 74,000 'overcrowded and unwillingly living in other people's households'.

Most of the estimates of the homeless population are under one per cent of the population. The figures cited for Germany (Table 4.4) are equivalent to 0.16 per cent of the total population as 'fully homeless' and 1.3 per cent when lesser degrees of homelessness are included. The estimates quoted by Schaefer for France are 0.4–0.8 per cent of the population. The recent estimates given in the US chapter range from 0.25 per cent to 1.3 per cent.

For the UK, the official figures of households accepted as homeless (under the Homeless Persons Act 1977) underestimate the absolute size of the problem but give a good indication of the trend. As Malpass points out, the numbers of households accepted annually more than doubled between 1978 and 1989, from 53,100 to 126,680. Another indicator of homelessness is the number of households publicly supported in temporary (mostly 'bed and breakfast') accommodation. This rose from 10,000 in 1970 to 40,000 in 1989, most of the increase taking place after 1982, when the sale of council houses and the cutbacks in building began to make an impact on the available lettings of council housing. However, these figures (even if trebled to give the numbers of persons) grossly underestimate the absolute size of the problem. Many people are rejected under the Act's provisions, or do not apply, since local authorities are obliged to accept only 'priority' groups, which usually means households with children or pregnant women. Consequently nearly all single people are excluded.

### Affordability problems

'Housing affordability', in the sense of difficulty in paying for housing, is a less dramatic issue than homelessness – except when it culminates in eviction or repossession. However, the numbers of households facing problems of this type are much greater. Howenstine gives estimates that 29 per cent of households in the USA faced a 'cost problem' in 1987. All the contributors conclude that the problems of housing affordability for the lower income groups increased during the 1980s. In the USA, the percentage of all households paying over 30 per cent of their income for housing costs rose from 14.7 per cent in 1975 to 23 per cent in 1987 (Table 2.8). However, the figure for very low-income owners rose from 27.2 to 48.2, while the figure for very low-income renters rose from 63.2 to 74 per cent. In West Germany, rent as a percentage of the income of the poorest household decile, after the housing allowance, rose from 21 to 32 (Figure 4.5). In France, rent/income ratio for all tenants, after the housing allowance, rose over the same period from 9.1 to 12.6 (Table 6.2) but the rent/income ratio for households on the minimum wage was 17 per cent and, for households not receiving the housing allowance, 33 per cent (Figure 6.1). For the UK, Malpass states that 'assured tenants' (of housing associations) were paying 24 per cent of their income in rents in March 1990.

This rising trend of rent/income ratios in the 1980s, although not in dispute, needs to be seen in historical perspective. The available statistics suggest that, in European countries between the two World Wars, rents averaged around one-tenth of income (Howenstine, 1986, chapter 3). This was substantially lower than before the First World War. In the USA, on the other hand, around 20 per cent appears to have been usual. After World War Two, Shelter Income Ratios (SIRs) were generally low in Europe, often as a result of rent control. In the 1970s, rents often lagged behind the high rates of inflation. Representative SIR percentages for the 1970s are given as 9.4–9.5 in France; 11.35 in the Netherlands; 11 in the United Kingdom; and 23 (renters) and 16 (owners) in the USA (Howenstine 1986, Table 2).

To some extent the rising rent/income ratios of the 1980s were part of a 'catching up' process after a period of often uneconomically low rents. On the other hand, the rises in rents (and the costs of owner-occupation) hit certain groups of households particularly hard. Thus Howenstine states that, in the USA between 1975 and 1988, the average real income of renters *fell* by 4 per cent, whereas rents, in real terms, rose by 17 per cent. Similarly, as a comparative Anglo-German study points out (Hills 1990), British local authority rents rose much faster between 1979 and 1988 than either housing association rents or private sector rents; thus many of the poorest tenants, concentrated in council housing, faced rent rises which exceeded the rise in their incomes. In the private rented sector in the UK, decontrolled rents were often double the level of controlled 'fair rents'. Young people seeking rented accommodation, who were usually dependant on the decontrolled private sector, therefore faced above-average rents. Ulbrich and Wullkopf mention a similar, although less pronounced, phenomenon in Germany.

Affordability problems, however, affected owner-occupiers as well as tenants and – especially in the UK and USA – owner-occupiers were not always affluent. Malpass draws attention to the hardship involved for many lower-income households who bought during the 'boom' of 1986–9, and cites estimates by Bramley (made before the fall in house prices and interest rates after 1990 but also before the accompanying depression) that 75 per cent of households in South-East England were unable to afford home-ownership. Howenstine refers to the 'home-owner's fading dream' in the USA. First-time home-buyers faced higher costs than home-owners in general, because people who bought some years earlier had their repayments eroded, in real terms, by inflation.

## The costs of ownership

In spite of the hardships faced by some house-buyers in recent years, it would be a mistake to think that housing costs for owner-occupation (in relation to income) have shown a relentless rise over the past two decades. Statistical time-series show a more complex pattern. Relatively good statistics on house prices are available for the USA and Great Britain. For the other countries, the statistics are more fragmentary, but a recent study has brought them together to produce fairly reliable time-series (Holmans 1991). One commonly used indicator of 'affordability' is the price/income ratio i.e. the average (or median) house price divided by average (or median) household income. Figures 8.1(a) and 8.2(a) show the price/income ratios for the USA and Great Britain, produced by the National Association of Realtors and the Building Societies Association. In Figures 8.3, 8.4, and 8.5, Holmans' house-price series have been deflated by the OECD figures of average hourly wages, as a surrogate for household income.

Price/income statistics of this type provide a reasonable indicator of long-term changes in affordability but, in the short run, changes in interest rates can have a marked effect on the level of mortgage repayments, which is what matters to the householder. More complicated estimates are needed to estimate the changes in repayments by a typical borrower, and are available only for the USA and Great Britain.

In Figure 8.1, for the USA, line (a) shows median house prices divided by the median household income; there was a steady rise in the 1970s to a level 25 per cent higher than in 1970. In the 1980s the ratio fell back, and stabilized at around 15 per cent above the 1970 level. The median price rose more in relation to the average wage, but median household income was boosted by the trend to two-earner families. Line (b) shows the estimated initial repayment on the purchase of a median-priced home, as a percentage of median household income. The percentage doubled in the 1970s, and then fell to around 35 per cent above the 1970 level. Thus, in spite of an improvement in affordability after 1982, the costs of owner-occupation over the 1980s as a whole were markedly higher than in the previous

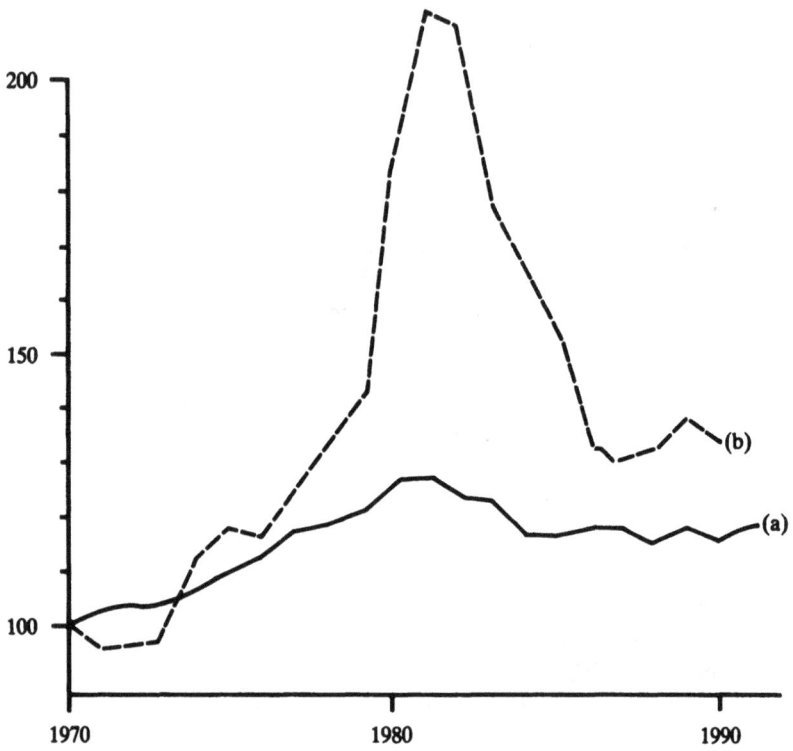

(a) —— Median price, existing homes, in relation to median household
income.
(b) ----- 'Affordability'. Estimated payment for median priced home, as
percentage of median household income.

*Figure 8.1*   USA – House price/earnings ratios and affordability index,
1970–90 (1970 = 100)
*Source:*   National Association of Realtors

two decades and, as Howenstine points out, the percentage of owner-occupation fell.

Figure 8.2 gives somewhat comparable time-series for Great Britain, using Building Societies Association data. Line (a) shows the average house price in relation to average earnings. The main impression is of the 'twin peaks' of 1973 and 1989, corresponding to the two great inflationary booms of the post-war period. House prices were depressed in the recession of the mid-1970s and in the depression (to use a perhaps more appropriate term) of the early 1980s; in between, there was an upward blip. With the onset of another depression in 1990 the price/income ratio fell sharply and, in the Spring of 1992, had fallen below the trend level (of around 3.5).

Line (b) shows an index of 'affordability' – the estimated initial repayment as a percentage of the average household income of first-time home-buyers. This index rose during the 'Barber boom' of the early 1970s (named after the presiding Chancellor of the Exchequer) and fell back in the subsequent slump. After 1978, however, the trend was sharply upwards. The ratio fell slightly during the first depression of the 1980s, but rose to unprecedented levels during the 'Lawson boom', peaking in 1990 at over double the 1970 figure. There followed a sharp fall (as a result of falls in house prices and interest rates); by the Spring of 1992, affordability was back at the level of the early 1980s.

The UK thus displayed exceptional volatility in house prices and affordability in the 1970s and 1980s. The major booms of 1972–4 and 1987–9 followed decontrol of the banks and building societies respectively, and were linked with inflationary upsurges. After 1989, house prices fell in nominal (i.e. money) terms, which had not happened in the previous cycles, when nominal prices had merely failed to keep up with inflation. This fall was part of a depression which, in the field of housing, was even more severe than its predecessor, and in which the number of repossessions rose sharply. Repossessions had not exceeded 3,000 a year before 1979 but reached 43,000 in 1990 and around 80,000 in 1991 and 1992. At the same time, low house prices and sales caused a sharp contraction of the housebuilding industry – which may be storing up supply problems for the late 1990s. The fall in house prices (which was concentrated in Southern England) was accompanied by a sluggish market in second-hand houses which inconvenienced potential movers and reduced the earnings of estate agents. This low level of sales was due partly to fears about employment, and partly to sellers still asking unrealistically high prices.

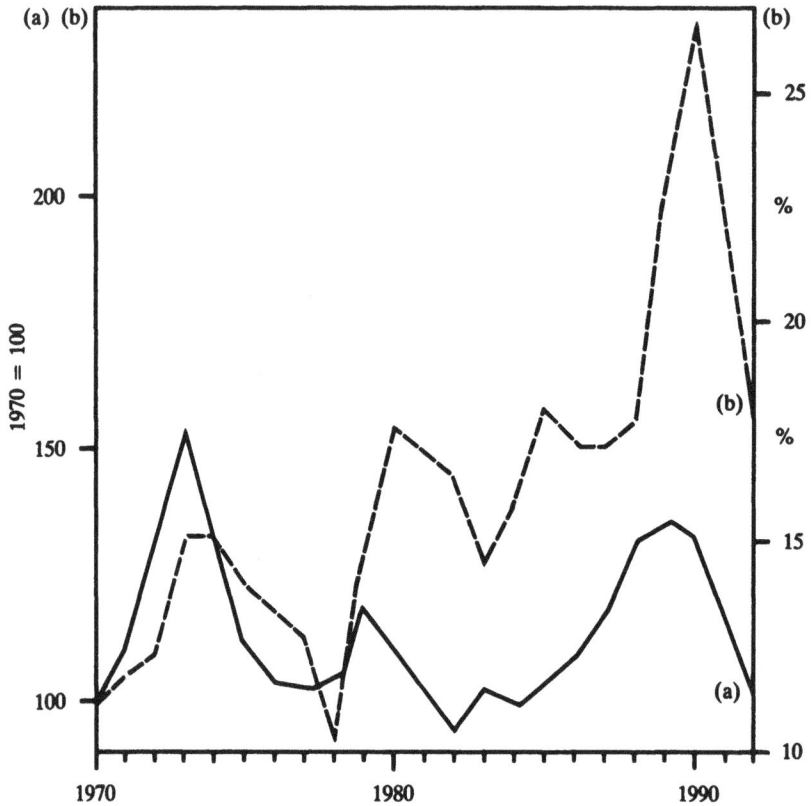

(a) ——— Average house price in relation to average earnings.
(b) ----- 'Affordability'. Initial repayment as a percentage of the average
income of first-time buyers.

*Figure 8.2*   Great Britain – House price/earnings ratios and affordability
index 1970–90
*Sources:*   (a) *Housing Finance* Aug 1991, p.13, Table 4. Building Societies
Association (b) *Housing Finance* Aug 1991, p.10, Table 1

The UK in the summer of 1992 was experiencing a 'cold turkey' cure for inflation, involving a fixed exchange rate in the European Monetary System. If a firm anti-inflationary policy is pursued for some years, inflation rates could stabilize at the German–French–Dutch level of (at least until the upward blip in 1991) 2–3 per cent, with fairly stable house prices and interest rates. Housing would be much more affordable than it was in the late-1980s, although unemployment would be likely to remain high, and it could take some years for repossessions to fall to the pre-1990 level. Such an economic environment would mean that a house would – as in continental Europe – be a place to live in rather than a hedge against inflation. However, it is perhaps too soon to celebrate the death of inflation and house price cycles. The Government's attempt to introduce a permanent low-inflation environment was coming under intense criticism from an alliance of monetarists and Keynesians, Right-wing Conservative and Left-wing Labour M.P.s, employers and trade unionists. Many of the critics not only called for devaluation, but rejected any limits on monetary expansion, and stressed the virtues of inflation. The *benefits* of lower house prices for housing affordability were hardly mentioned; the whole stress was on the difficulties faced by housebuilders, and the effect of low house prices in restraining consumer expenditure. British attitudes to inflation and the housing market still seemed to be very different from those in Germany, France and the Netherlands. (PS. Devaluation occurred on 17 September 1992. A London estate agent (realtor) expressed frankly the reaction conveyed in coded language by the spokespersons of estate agents, housebuilders and mortgage lenders. 'Mr – said he welcomed the prospect of higher inflation as a result of a weakening pound. "The house market will not get going unless there is inflation"' *Financial Times*, 19 September, p. 4.)

### The continental European countries

For France, only fragmentary house price series are available (Figure 8.3). Line (a), representing average house price/income ratios for three-year periods from 1967–70 to 1981–4 suggests that there was a decline from the late 1960s to the late 1970s. Thereafter, there was probably stability followed by a rise; line (b) suggests that the price/income ratios dipped in the depression of the early 1980s, but began to climb after 1984. Line (c) indicates a rise of 20 per cent between 1987 and 1990. Although the statistics are not altogether satisfactory, they suggest that the price/income ratio has displayed a

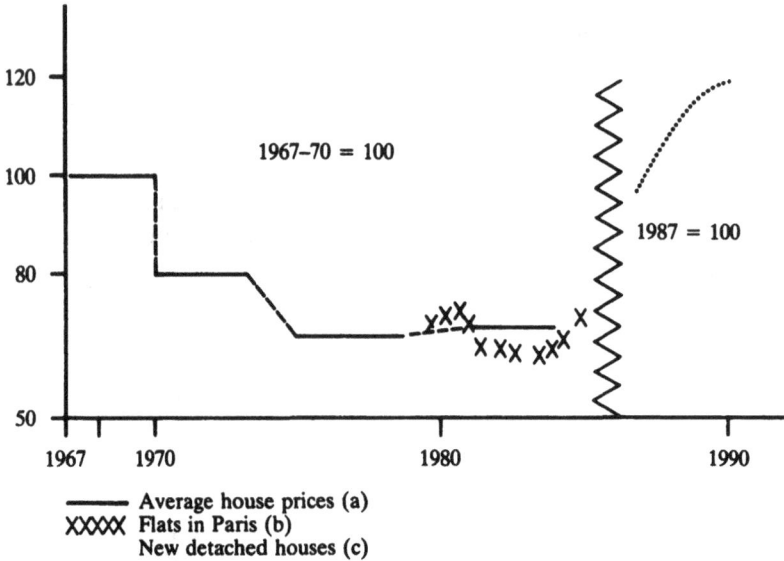

Figure 8.3   France – house prices in relation to average wages 1967–90
*Source:*   Holmans (1991) (a) Table B1; (b) Table B5; (c) Table B6, OECD
*Main Economic Indicators*

slight fall-and-rise since the early 1970s, but relative stability. There
were considerable regional variations in house prices; in the late-
1980s house prices in Paris (city) were between 2 and 2.5 times
higher than in urban areas outside the Paris conurbation, and some
80 per cent higher than in the Paris suburbs (Holmans 1991, Table
B4). Schaefer points out that, in spite of relative house price stability,
the high interest rates of the early 1980s caused difficulties for some
buyers, although these were alleviated by a scheme for renegotiating
(fixed interest) mortgage loans. Neither France nor Germany or the
Netherlands has, however, experienced the level of debt-induced
problems experienced in the UK.

German house price/income ratios have been relatively stable, with
a downward trend since 1972, in spite of the sharp rise after 1989
(Figure 8.4). A considerable divide in house prices has emerged over
the past twenty years between North and South, or rather between
the three large Southern cities of Frankfurt, Stuttgart and Munich,
and the rest (Holmans 1991, Chapter 3).

The Netherlands experienced a rise in the house price/income ratio

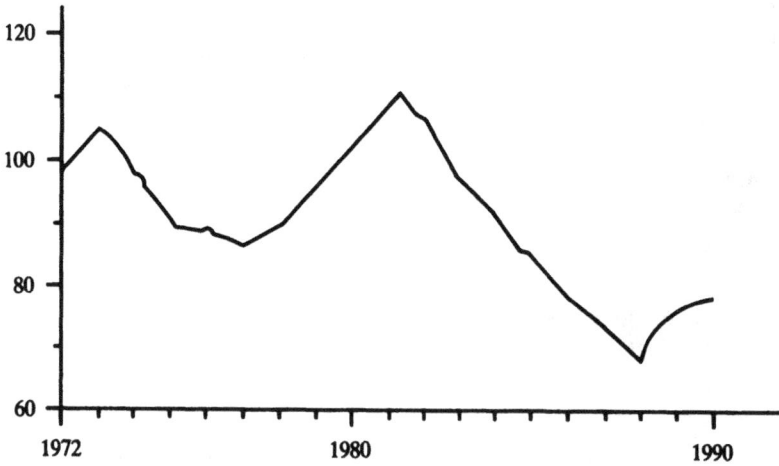

*Figure 8.4* West Germany – house prices in relation to average wages
1972–90 (1972 = 100)
*Source:* Holmans (1991) Table C1; OECD *Main Economic Indicators*

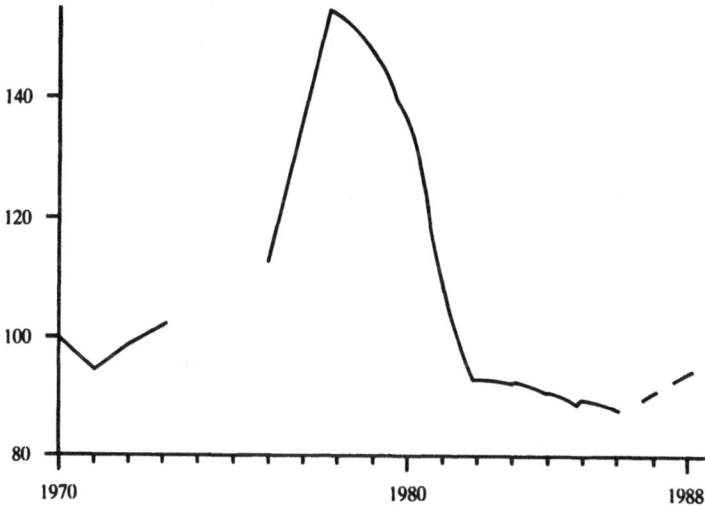

*Figure 8.5* Netherlands – house prices in relation to average wages 1970–85
(1970 = 100)
*Source:* Holmans (1991), OECD *Main Economic Indicators* Table D1

of about 50 per cent from around 1975 to 1978, followed by a fall of
a similar amount between 1979 and 1982 (Figure 8.5). The Dutch
Government stopped collecting price statistics after 1985, but the
indications are that the ratio has since remained fairly constant. The

*Table 8.2* House price/income ratios: characteristics of national housing markets, mid-1980s

|  | USA | France | Germany | Netherlands | UK |
|---|---|---|---|---|---|
| House price/income ratios* | | | | | |
| a new houses | 6.7 | 6.2 | – | – | 6.9 |
| b second-hand houses | 5.9 | – | 9.6 | 5.0 | 5.8 |
| First-time buyers as % of all buyers | 37 | 75 | 75 | 65 | 37 |
| Second-hand dwellings as % of dwellings sold | 84 | 50 | 35 | 50 | 87 |
| Dwellings sold second-hand as % of owner-occupied stock | 6.7 | 4 | 2 | 3 | 9 |

*Source:* Holmans, 1991; Tables 1, 6
* average 1984 & 1988; median/average house price ÷ GDP per capita

rise in the late 1970s was part of a general inflationary upsurge, and the fall was a product of the painful 'stabilization crisis' which, as in France, brought inflation down to the German level.

### Characteristics of national housing markets

The changes in house prices and affordability outlined above have to be seen in the light of differences in the structure of national housing markets (Table 8.2).

The figures on house sales indicate a sharp difference between the USA and UK, on the one hand, and France, Germany and the Netherlands on the other. In the 'Anglo-Saxon' countries, the housing market was dominated by second-hand houses, which accounted for 84 per cent of sales in the USA and no less than 87 per cent in the UK (although Americans are more geographically mobile). In the continental European countries, second-hand houses accounted for 50 per cent of the market in France and the Netherlands, and only 35 per cent in Germany. This reflects the fact that households become owner-occupiers later in life than in the Anglo-Saxon countries, and then move less often, partly because owner-occupation is not (yet) the norm.

House price/income ratios in the UK around 1985 were similar to those in the USA, slightly higher than in France or the Netherlands, and substantially lower than in Germany. The houses involved, however, may not be comparable. The quality of German houses is undoubtedly high, but the German price/income ratio seems out of line, and reinforces the conclusion of Ulbrich and Wullkopf that home-ownership is virtually impossible for households on medium-to-low incomes, unless they inherit wealth.

## THE CAUSES OF HOMELESSNESS

What were the causes of the increase in homelessness and of the affordability problems faced by tenants and home-buyers – especially lower income ones – in the 1980s? The issues, although related, can best be examined separately.

There are (see the US chapter) three main types of homelessness: 'economic', 'situational' and 'chronic'. People in the first group simply cannot afford housing, so that they will cease to be homeless if they are provided with either housing they can afford or the necessary cash to pay prevailing rents. The second group are people facing some kind of short-term crisis – young people who have left home precipitately, people experiencing temporary psychological problems, etc. – who may need training or social help as well as temporary accommodation. The third group are the chronically mentally handicapped, or mentally ill, who need permanent housing with supervision.

In the USA, as Howenstine points out, shelters provide a temporary but not a permanent, solution for homeless people. The numbers of these shelters – mainly run by voluntary bodies, with public financial assistance – increased several-fold in the 1980s. Surveys suggest that around a third of US shelter-dwellers suffer from severe psychiatric problems and about 40 per cent are drug takers (Economist 1991). The National Alliance for the Mentally Ill estimates that, of the Americans with very disturbed mental conditions, only 70,000 remain in psychiatric hospitals, but at least 150,000 live in shelters or on the streets, while 100,000 are in prison. However, the policy of closing mental hospitals began in the 1950s, but the numbers of homeless people did not begin to rise sharply until the 1980s. Nor can this rise be attributed to cuts in public housing. As Howenstine shows, the USA had very little public housing to begin with but, on a broad definition including Section 8 housing, the amount was (thanks to Congress) not significantly reduced during the Reagan era.

Other changes can, however, help to explain the rise in homelessness in the 1980s – most notably the decline in the supply of low-rent private rented housing documented by Howenstine. For many years, most American cities had 'Skid rows' – insalubrious poor areas providing cheap accommodation in 'houses in multiple occupation' and single-room-occupancy hotels. In the 1980s, many of these areas were redeveloped for commercial or high-income residential uses.

In the UK, the number of mentally ill people among the homeless in Britain is reported to be growing. The closure of mental hospitals began later in Britain than in the USA and, under a central Government programme, creditable attempts have been made to provide 'ordinary homes' for the ex-inmates. Nevertheless, there have been cases in which mental hospital patients have been put out on the street, and there may well be more, when the 'Care in the Community' programme is handed over in 1993 to local authorities which are financially broken-backed and in the throes of yet another reorganization. West Germany and France have adopted a more conservative policy towards the closure of mental hospitals, but similar problems are likely to emerge before long.

It is sometimes argued – in the UK as in the USA – that homelessness is 'voluntary', in the sense that homeless people prefer the life, or that young people leave home fecklessly. The available British studies indicate that this is not the case. Young people are leaving home earlier, but they are often leaving intolerable family conditions. One survey of (mostly young) people staying at a night shelter in London indicated some of the causes of begging and homelessness (Economist 1990):

1 Teenagers have had their entitlement to social security cut. Many homeless people are recent school-leavers who have left home, and have to beg to eat.
2 Social security payments, for those entitled to them, are more difficult to obtain. The old 'of no fixed address' allowance, collected daily, has been abolished, and most benefits are payable two weeks in arrears.
3 More young people are being pushed onto the street from institutions. Forty-one per cent of the shelter clients had been in childrens' homes, compared with 23 per cent in 1987. Two-thirds said that they could not return to their last settled base.

The extent to which the current British social security 'safety net' fails to help most homeless people was dramatically illustrated in a 'World in Action' TV programme in March 1992, in which a 'homeless' man with a camera bag had to spend a month in London. His treks from one social security office to another in a fruitless attempt to obtain the few pounds to which he was entitled were truly Kafkaesque; he spent most nights sleeping rough. Investigative TV programmes of this type have a unique ability to cast light into dark corners.

It is clear that 'homelessness' covers a wide range of situations. However, the fact that many homeless people face problems other than housing does not mean that the cost and availability of housing is irrelevant. One review of homelessness in Britain concludes that: 'The market gap between low incomes and the price of housing remains an essential feature of homelessness' (Greve and Currie 1990 p. 4). For some homeless people, the availability of moderately-priced housing would be enough; for others it is one of several requirements. There is also a vicious circle. Being homeless itself encourages psychological illness. At the very least, it gives a person an unkempt appearance which reduces still further his/her chance of obtaining employment, and so obtaining an income which might make it possible to obtain housing. Indeed, most employers will not employ someone without an address.

The number of 'hidden homeless' – as indicated above – greatly exceeds the number of 'roofless', and it is easy to slip from one situation into the other. Even for people in normal tenures, a change of circumstances can lead to sudden and unexpected homelessness. The main immediate causes in Britain are the breakdown of sharing arrangements or marriages/partnerships, or the loss of a private tenancy. However, as Malpass points out, inability to continue paying the mortgage is becoming increasingly important. Even 'priority cases' such as families with children   which are in theory covered by the Homeless Persons' Act – can find that they have unwittingly excluded themselves from the Act's provisions. For example, a family facing repossession which vacates the property voluntarily, rather than waiting to have an eviction order served on them by the courts, is likely to be deemed to have 'deliberately made themselves homeless'. This seemingly unexceptional clause has even been used – by overwhelmed local authorities – to reject families with young children who have been evicted because of rent or mortgage arrears. There is anecdotal evidence – as in the USA – of a growing number of families (social workers estimate that it might be around 250,000) 'doubling up', often with parents, in grossly overcrowded conditions. In one respect, however, there is a difference between, on the one hand, the USA and, on the other hand, the UK and other European countries. In the USA, as Howenstine points out, a severe illness for any member of the family (under sixty-five), requiring major medical treatment, can sometimes lead to impoverishment and even homelessness. The national health services of the European countries have removed that risk.

## THE CAUSES OF AFFORDABILITY PROBLEMS

A commodity can become 'unaffordable' for households if (a) the households' income falls, (b) the price of the commodity rises or (c) cheap supplies of the commodity cease to be available. There is evidence that all these things happened to housing in the 1980s. In most of the countries there was a growth of relative poverty i.e. the incomes of the poorer households (at least in some social groups) rose less than the average, even fell. The rents of social housing rose as a result of a reduction in subsidies or, in some countries, the selling-off or deregulation of social housing. Rents in the private sector rose because of a relaxation of rent controls, or because of a reduction in the supply of low-rent housing for commercial reasons. House prices and interest rates rose – at least in certain periods.

At the same time, the demand for housing increased as a result of social changes: people formed households earlier, and broke them up more often; life expectancy increased. The effect was to produce a steady fall in household size and 'underoccupation'; this is documented in the chapter on France, but applies to all the countries studied. In some countries, notably Germany and the USA, there has been large-scale immigration. The rise in rents (and in some countries the cost of home-ownership) was offset by increased housing allowances – but not always completely. Thus, although the mix of factors varied from country to country, most countries displayed: (a) increased poverty (b) an increased demand for housing from new, often not very affluent, types of household: (c) less, or more expensive, social housing, and a generally less sympathetic government attitude to people with housing problems: (d) a tendency for the supply of low-rent private housing to decline. Further factors in the UK and USA were: (e) the 'debt crisis' and, possibly, (f) land policies.

### The growth of poverty

The evening-up of incomes which characterized most Western countries for the three decades after the Second World War ceased in the 1980s and, in some countries, went into reverse. This change (related to the growth of unemployment in the European countries) has had a major impact on both homelessness and on affordability problems in normal tenures.

The levels of relative poverty in the USA and European countries, and the changes between 1979 and 1986, have been the subject of a recent report by the Joint Center for Political and Economic Studies (McFate 1991 p. 1) which concludes that:

> Poverty and inequality increased in the 1980s in most Western democracies . . . . National economic restructuring, the result of global trade and production, has left a large proportion of working-age citizens in industrialised nations unemployed, underemployed and economically insecure. . . . Although the United States experienced more steady growth and lower unemployment than most of Europe in the 1980s, the U.S. poverty rate was double that of every continental European country by mid-decade.

Poverty was measured in terms of income per household, adjusted for family size, after taxes and transfers. Households with less than 50 and 40 per cent of the median household income were classed as being respectively in 'poverty' or 'more severe poverty'. On both the 50 and 40 per cent definitions, the USA had the highest poverty rates in all categories of household (Table 8.3). These calculations are in line with considerable descriptive evidence of extensive (albeit relative) poverty in the USA, concentrated in a largely (but by no means wholly) black or hispanic underclass. The European countries, however, are experiencing growing problems of a similar type, even if so far on a smaller scale. Moreover, the whole of Western Europe now faces, in Eastern Europe, a counterpart to the 'huddled masses' of the USA's Southern 'back yard'. If economic conditions in Eastern Europe do not improve, illegal immigration could become as significant a factor in Western Europe as in the USA.

*Table 8.3* Non-elderly (1) households in 'poverty'* or 'more severe poverty'** – 1979 and 1986

| | All households* | | All households** | | Couples with children* All (1) | | Younger (2) | | Single parents* |
|---|---|---|---|---|---|---|---|---|---|
| | 1979 | 1986 | 1979 | 1986 | 1979 | 1986 | 1979 | 1986 | 1986 |
| | % | | % | | % | | % | | % |
| USA | 15.6 | 18.1 | 10.8 | 13.6 | 20.2 | 23.8 | 27.2 | 39.5 | 53.3 |
| UK | 8.5 | 12.5 | 3.6 | 7.0 | 10.6 | 16.8 | 14.2 | 23.2 | 18.0 |
| W. Germany | 5.5 | 6.8 | 2.1 | 3.2 | 6.0 | 7.9 | 6.8 | 18.8 | 25.5 |
| Netherlands | 7.7 | 7.6 | 6.0 | 5.7 | 6.5 | 7.3 | 4.3 | 13.1 | 7.5 |
| France | 9.9 | 9.9 | 5.4 | 6.1 | 10.4 | 10.7 | 10.4 | 8.2 | 9.1 |

* Less than 50 per cent of adjusted median household income
** Less than 40 per cent of adjusted median household income
(1) Aged 20 to 55 (2) Aged 20 to 29

*Source: Poverty, Inequality and the Crisis of Social Policy*, Joint Center for Political and Economic Studies, Washington, D.C., 1991, Katherine McFate: Luxembourg Income Study

It should be added that the official US 'poverty line' figures, covering all people under sixty-five, paint a somewhat brighter picture than the Joint Center's figures of relative poverty for people under fifty-five. However, the 'President's National Urban Policy Report' (US 1991), although generally upbeat, does not make any great claims for the achievements of the 1980s:

> The war on poverty among persons under 65 – the poor of working age and their families – has been at best a draw during the 1980s, and then only as a result of the strong economic expansion that began in late 1982. (When properly adjusted for inflation, the poverty rate among this group declined from 14 percent in 1983 to 11.7 percent in 1989.)

The Joint Center's figures indicate that, between 1979 and 1986, the UK moved away from the 'European' towards the 'US' level of relative poverty. In 1979, the British poverty rates on the two definitions (8.5 and 3.6 per cent) were broadly in line with those in the continental European countries. By 1986, the rates in Germany had risen slightly, while those in France and the Netherlands were the same or slightly lower. But the British rates had jumped to 12.5 and 7.0 per cent. These were still well below the rates for the US, but well above those in the continental European countries. Table 8.2 also shows that, in all countries, the rise in poverty was particularly sharp for couples (especially younger couples) with children. Poverty is becoming increasingly a problem of youth rather than of old age.

The growth of poverty in the 1980s was largely a result of a greater inequality in pre-tax incomes. In other words, it was a reflection of deep-seated changes in economic conditions, which were strikingly similar in all the Western countries studied. However, more regressive tax/benefit systems also played a role in the UK and USA. Welfare (social security) payments for many low-income groups were cut, or made more difficult to obtain; this particularly affected young people, and homeless people.

The combination of falling incomes and rising housing costs produced a 'scissors' squeeze on the poorer social groups. This happened in most of the countries, but is well illustrated by some illuminating official statistics for the UK. The share of total net income accruing to the bottom fifth of the income distribution fell from 10.1 in 1979 to 8.9 per cent in 1987 (Central Statistical Office 1991). (These are the revised figures; the highly publicized original figures showed no fall in the percentage, but were discovered by an independent researcher to contain 'computational errors'!). The shares of the next three-fifths also fell, whereas the share of the richest fifth rose from 34.3 to 39.1 per cent. When net income *after housing costs* is calculated, the differences between the top and the bottom of the income distribution become even greater; the share of the bottom fifth fell from 9.5 to 7.6 per cent, while the share of the top fifth rose from 35 to 41 per cent. The differences in the figures before and after housing costs indicate that, over the period, housing costs rose more sharply for the poor than for the rich. A subsequent survey concluded that the bottom 10 per cent of households experienced an *absolute fall* of 6 per cent in real disposable income between 1979 and 1989 (CSO 1992). In short, the poor got poorer, and at the same time faced above average rises in housing costs.

### Social changes affecting housing

In addition to these changes in income distribution, there have been social changes – common to all the countries studied, if in varying degrees – which have affected the housing situation. In the chapter on the Netherlands, van Weesep and van Kempen express the changes as follows: 'Secularization and the seeking of self-fulfilment – the desire to realize more of one's own potentials – underlie the break with traditional values'. The '"me, now" generation' is a less flattering description. In any event, young people want to leave home earlier, and the numbers of divorced people and single parents have risen sharply. At the same time the number of second homes has risen, and the supply of cheap rented housing has tended to fall, as households become less willing to let surplus rooms in their houses, and landlords move self-contained rented properties 'up market'.

### The decline of social housing

In his chapter on the UK, Malpass attributes much of the blame for the 'housing crisis' to the policies of the Conservative Government towards council housing. Ulbrich and Wullkopf are equally critical of the accelerated deregulation of social housing (which, however, began to have an impact only towards the end of the 1980s). In the USA, the Reagan Administration was also hostile to 'public' housing, but the situation was different from that in Europe. As Howenstine points out, public housing in the USA started out under the control of unsympathetic local interests, and with the ethos that it should be the cheapest and most basic type of housing, often on out-of-the-way sites, designed for people who were poor, but 'deserving'. Tenants who exceeded the income limits for eligibility had by law to be evicted, which was disruptive of a stable community. Public housing in the USA also suffered from the well-publicised consequences, in some inner-city projects, of a concentration of poverty and crime. Most public housing was outside the 'problem' districts but, in Howenstine's words, 'the stereotype became so entrenched in political dialogue that it became difficult to maintain popular support for the public housing programme'.

The situation on the other side of the Atlantic was very different. In France, West Germany and the Netherlands a substantial proportion of social housing tenants came to have incomes over the qualifying limits. This situation contributed to a relatively low level of 'stigmatization', although it gave rise to controversy. In the UK, no national income limits were laid down, but local authorities took income into account in accepting tenants. Although British council housing was always more segregated than social housing on the Continent, it was, until recently, judged favourably by independent observers. One American, writing in the early 1970s, stated:

> UK council housing is, on the whole, of good quality, well maintained, and not overwhelmed by vandalism and crime. It is on a sound financial basis. It has a socio-economic mix and is not all bunched together in the poorest or high minority concentration neighborhoods. And it is accepted as legitimate by the vast majority of the British population, and most particularly the working class. Each of these attributes stands in marked contrast to public housing in the United States.
>
> (Wolman 1975)

In the 1980s, British council housing lost many of the attributes praised by Wolman. The reasons for the change are complex. As a government economist has pointed out, the fact that many political criticisms of council housing were exaggerated should not obscure the fact that it lost the broad political support which had previously existed (Holmans 1991). Peter Malpass points to one underlying cause: the change in the social composition of tenants as the 'artisans' moved out, and poorer groups moved in from the declining private rented sector. This change deprived council tenants of effective political leaders, and probably contributed to the physical deterioration of many council estates – which probably cannot be attributed solely to poor design and cut-backs in maintenance.

British council housing, like US public housing, also suffered from untypical stereotypes, often embodied in memorable pictures – the dynamiting of the Pruit-Igoe project in St. Louis; the box-of-cards collapse of one side of the Ronan Point tower block in East London. However, although most council housing was unlike Ronan Point (being in fact 'houses'), Ronan Point was typical of the type of housing (tower blocks, long corridors, unsupervised space, untried 'systems building') which – often under pressure from Whitehall – was constructed in the 1960s. This housing soon proved to be ideal for thieves and muggers, and also liable to rapid physical deterioration. Another factor in the decay of some estates was sheer size. Until the 1950s, the numbers of dwellings administered by councils rarely exceeded a few thousand. The management of estates of over 50,000 or 100,000 dwellings presented many council housing departments with a managerial challenge which they were unable to meet.

These failings of council housing provided targets for a radical privatization programme – the 'right to buy' (at discounts of up to 60 per cent on market value) introduced by the 1979 Conservative Government. This programme, under which 1.4 million dwellings have been sold, has no parallels in other countries. In France and the Netherlands, sales in recent years have been less than 0.5 per cent of the stock, and confined to schemes for diversifying and modernizing large estates. The West German privatization is different, with social housing on which loans are repaid becoming mainly privately rented.

The overt case put forward for the British 'right to buy' was that it would provide families with the power to remodel their homes as they saw fit, enjoy the long-term security of home-ownership, and counter the 'dependency culture' which was allegedly fostered by council housing. No large-scale study of the economic and social consequences of the 'right to buy' has been undertaken, but it seems clear that many former tenants have benefited. (The fact that the 'right to buy' has now been endorsed by both the Labour Party and the Liberal Democrats speaks for itself). On the other hand, there has been a fall in the number of dwellings available for letting to new tenants. Since the demand for social housing has not fallen correspondingly, the consequence (as Malpass points out) has been considerable excess demand for social housing in most parts of the country. The majority of lettings are now to homeless families, and council housing is becoming an exercise in crisis management.

The 'right to buy' also had two unfortunate consequences on the composition of lettings. The more attractive properties were naturally bought first – especially houses as distinct from flats. Flats made up 30 per cent of the council stock, but have made up only 5 per cent of sales. The councils have been left with the less attractive properties, especially the sometimes structurally unsound non-traditional housing of the 1960s. The concentration of increasingly poor tenants in poor housing has accentuated the problems of 'ghettoization' which characterized much US 'public housing'. Second, the results of the 'right to buy' have varied geographically. In some districts – notably the more desirable parts of London and rural districts – council housing has been virtually eliminated. In many cities there is still a significant amount of council housing, which is likely to remain social housing, even if councils cease to be the landlord. However, in rural areas (which in most parts of the UK, in contrast to France or the USA, have high house prices) the small numbers of council houses were often the only housing that local working-class people could afford. When these houses were sold, affordable housing virtually vanished. Thus the 'right to buy', combined with the purchasing power of commuters and second-home buyers, has produced a shortage of affordable housing in rural areas (especially in Southern England) which is small in absolute numbers but even more acute than in the cities.

These criticisms of the 'right to buy' do not mean that social housing should never have been sold. There were some British cities in which very little except council housing had been built for decades, and where a reduction in the dominant position of council housing could certainly be justified. In most cities, some council tenants were in a position to become owners, even without rebates, while still leaving a substantial stock of social housing. Moreover, the nature of demand for social housing changes over time; it can make sense to sell housing of certain types and in certain districts in order to invest in other types and other districts. British councils always had the right to sell their housing, and some made use of it in this way. The 'right to buy', by contrast, allowed no discretion to councils or even non-charitable housing associations. Moreover, it was accompanied by Government edicts which prevented councils from spending the proceeds of sales on new housing. Thus in the Spring of 1992, British local authorities, many of them with acute problems of homelessness and housing shortage, had £8 billion standing idle in their housing revenue accounts (enough to buy about 160,000 houses), which they were not allowed to spend on housing.

### Social housing in continental Europe

The continental European countries also experienced some difficulties with social housing, but the real, or alleged, problems were not as great as in the UK and USA, and the moves to privatize social housing were not so strong. In France, the *grands ensembles* suffered from problems of size and monolithic design as well as from problems arising from a concentration of North African immigrants. West Germany's architectural problems were on a smaller scale, partly because most social housing was undertaken by relatively small enterprises (*Neue Heimat* being an exception). West German social housing did, however, have an economic problem, deriving from the fact that the rents were based on the historic costs of individual buildings. Over time, this produced excessive differences in rents between older and new dwellings. The rents of new social housing were sometimes above those prevailing in the private sector. (In the UK, this difficulty was avoided by allowing councils to pool their rent receipts.) The situation in the German Democratic Republic was very different from that in West Germany. The large housing estates contained a virtual cross-section of society and had few 'social problems'; on the other hand, the defects of remote and bureaucratic management were present on a massive scale.

The Netherlands built up, and has retained, a very large social housing sector. Although free from 'stigma', and generally well managed, it has also had its problems. Van Weesep and van Kempen emphasize the 'misallocation' of low-rent housing. This is a magnified version of the West German problem, and is one which people in the UK or USA may find difficult to grasp. It arises from the fact that the elaborate and extensive system of subsidized housing, with rents based on the historic costs of individual dwellings, has produced a pattern of rents which has little relation to current values. This might not matter if poor people occupied low-rent dwellings and affluent people occupied higher-rent dwellings. In practice, such a tidy allocation cannot be maintained, and it is disruptive of communities to ask tenants to leave if their income rises above the qualifying level for social housing. France and West Germany have had similar problems (*Fehlbelegung*) and have resolved them, at least partially, by means of rent surcharges, but such a solution has proved politically unacceptable in the Netherlands.

This *impasse* leads van Weesep and van Kempen to call for greater freedom to build unsubsidized high-rent and owner-occupied housing in the cities, so that 'filtering' can make more low-rent accommodation available for lower income households. This prescription might at first sight seem surprising, coming from supporters of Dutch 'social democratic' ideals. The encouragement of filtering (or 'trickle down'), on the grounds that the best way of helping the poor is to help the rich, has been criticized (especially in the USA, where it has been most influential) on the ground that it may just help the rich. However, the situation in the Netherlands is very different from that in the USA. The public sector has dominated new construction in the major Dutch cities, supplying up to 90 per cent; this has led to a very selective population structure and an unsatisfied demand for owner-occupied and luxury rental housing. While urging that this demand should be satisfied, and that the private sector should play a larger role, van Weesep and van Kempen support an active government policy for protecting the interests of the poor – in particular by maintaining a substantial stock of social housing.

It should be clear from the experience with social/public housing in the various countries that it cannot be regarded as a 'good thing' or a 'bad thing'. In all countries, it has made a valuable contribution towards the provision of affordable housing. Problems can arise, however, if it is restricted to very poor households and 'stigmatized'; if it is managed in large and bureaucratic units; if anomalies arise in relative rents; or if the sector is expanded to such a degree that the private provision of housing is inhibited. Conversely, radical attempts to privatize social housing can cause hardship for vulnerable social groups.

**The debt crisis**

Large-scale mortgage foreclosures have come to be an item on the political agenda in recent years, most notably in the UK, in 1991, but to a lesser extent in the USA, France and the Netherlands earlier in the 1980s. The problem arises because of a house price boom followed by recession. Households buy when prices are high, taking on mortgage repayments which they can only just afford, and then suffer a loss of income in the subsequent slump; they can easily get into arrears and face repossession. They may even find that the mortgage loan exceeds the market value of their house which (in Britain and the Netherlands) places them in an impossible position. The USA has a more borrower-friendly provision; mortgage lenders can repossess the house, but cannot claim any further outstanding debt. A similar provision, and various legal restraints on repossession, have been proposed by the Law Commission (1991). The 'information society' has brought a new hazard. Mortgage lenders now record repossessions and arrears on central computer systems and, in the UK, anyone who has been repossessed is unlikely to obtain a mortgage ever again. In the USA, there seems to be a greater willingness to wipe the slate clean after some years.

The new phenomenon of 'negative equity' (i.e. the situation in which the market value of the house is less than the outstanding loan) raises problems even for households who are able to continue making the mortgage payments. They may find it impossible to move house, even if they could sell their house, which – in the (no doubt temporarily) very sluggish market – may be difficult. The Bank of England estimates that nearly a million households are in this situation (Bank of England 1992).

In December, 1991, the Government, realizing that the repossession crisis could have electoral implications, pressed the building societies to 'do something'; the inducement offered was that social security payments for mortgage payments (which are available to some unemployed households) would be paid directly to the societies. The societies agreed that houses owned by households facing repossession would, in some cases, be temporarily transferred to housing associations, who would charge the occupiers a rent. (This would have the additional advantage of giving the occupants an entitlement to housing benefit.) This scheme was devised after two days' discussion with the building societies, and none with the housing association; not surprisingly, many questions remained unanswered.

A study of the first six months' operation of the scheme concluded that it had been a failure (Joseph Rowntree Foundation 1992). The payment of the mortgage interest element in 'income support' direct to lenders had little effect, because four-fifths of the recipients were already passing over the amount in full. 'Mortgage-to-rent' schemes were not popular, since most mortgage holders wished to remain home-owners. The most useful thing the mortgage lenders could do was simply to allow debtors more time, but the most effective scheme would be a 'mortgage benefit' (see p. 251).

Underlying these short-term efforts to help dispossessed households is a grave financial crisis in the UK and the USA (and Japan), inherited from the boom of the late 1980s. The crisis has its roots in housing and commercial property, and can best be explained by the theory of 'debt deflation', put forward in the 1930s, and subsequently forgotten (Fisher 1933). The theory is that a credit boom leads to property values being bid up to unrealistic levels. A slump in property values then occurs, which bankrupts some financial institutions and individuals, and induces others to cut back sharply on lending, leading to a recession.

After the property boom-and-slump of 1973–5 (in several countries) it was generally assumed that borrowers and lenders had learned their lesson, but the credit boom of the late 1980s in the Anglo-Saxon countries exceeded, in the extent of imprudent lending and outright fraud (the S&Ls, Maxwell etc), anything seen in 1973–5. In the UK, the problem of housing debt was aggravated by government pressure on council tenants to become owner-occupiers, and by the shortage of private rented housing. Germany, France and the Netherlands did not experience such an expansion of unwise credit in the 1980s, and had a larger supply of private rented housing. A further complicating factor in the UK is that mortgage lenders 'borrow short and lend long', rather than issuing long-term debentures. Thus repayments depend on volatile short-term interest rates.

In the past, problems of excessive total borrowing have often been resolved by allowing inflation to erode the value of outstanding debt. This was the policy implicitly adopted by the UK and the USA in the 1960s and 1970s. If this option has been ruled out (which is far from clear), bringing the debts incurred in the 1980s down to manageable levels could be a protracted and painful process.

**Land policies**

A final issue concerns the extent to which a shortage of land has restricted the supply of housing and forced up its price. The British town planning system was criticized by building firms during the house-price booms of 1972–4 and 1986–9, and these criticisms were made the basis of general theories by some academics. The dramatic falls in house and land prices since 1989 have cast considerable doubt on these theories. On the other hand, Britain's consistent failure to devise the type of public/private partnership in land policy found in the other European countries may well have had a prejudicial effect on the supply of low-cost housing (Hallett 1988). In the USA, 'growth management' (often hard to distinguish from 'exclusionary zoning') seems to have raised house prices in the late 1970s, although there was a subsequent reaction against these policies. The 'density bonuses' mentioned by Howenstine would not be necessary under a zoning system which made provision for low-cost housing.

Land policy in the continental European countries is less controversial. The German Federal Republic has a system for the acquisition of building land in which land banking and forward planning by local authorities has played a major role. Despite some criticisms, the system is generally considered to have worked well in terms of settlement planning and land supply for housing (Hallett 1977). In the 1980s, however, when it was universally believed that only a relatively low level of housebuilding would be needed for the foreseeable future, local authorities cut back their activities considerably, and the legislative basis for major settlement planning was repealed (Hallett and Williams, in Hallett 1988). With the sudden influx of population after 1989, it was difficult to reorientate policy sufficiently rapidly, especially in view of growing 'ecological' restrictions on development. There has been a sharp increase in land prices in the conurbations since 1990.

The Netherlands, as a result of its geography, has long had a system in which land acquisition is virtually a local authority monopoly. According to one expert, the system has eliminated the problems encountered in other countries (Needham, in Hallett 1988). In his discussion of current affordability problems, van Weesep does not discuss the aggregate supply of building land, which does not appear to be a significant factor.

France has a land acquisition system in which public bodies play an important, although often indirect, role (Pearsall, in Hallett 1988). Schaefer mentions the relatively high level of housing production in France, which suggests that land bottlenecks have not held back the supply of housing. As in the Netherlands, affordability problems are connected more with access to housing for low-income groups than with the aggregate supply of housing.


## WAYS OF RESOLVING THE 'NEW HOUSING SHORTAGE'

One way of 'solving' a problem, which governments are often tempted to adopt, is to suppress the evidence that it exists. The pre-eminent exponent of this policy was the GDR, but the US Federal Government and the British Government made some determined efforts in this direction in the 1980s. Howenstine documents the dismantling and corruption of HUD. In the UK, research on issues such as homelessness was cut back, and statistical services were, as one review recommended, 'tailored more closely to ministerial priorities and the political philosophy of the government' (Newson and Potter 1985) by being brought under the direct control of the Chancellor of the Exchequer; they were also weakened by staff cuts. 'Computational errors' by the Central Statistical Office (which were probably just that) attracted so much criticism, however, that the Government was forced in 1992 to undertake a managerial reorganization to restore the technical competence of the statistical service.

**Policies towards homelessness**

In the 1980s, public policies towards homelessness in the UK and USA were markedly unsympathetic. The Department of the Environment (which is responsible for housing) declined to send a representative when the professional body representing housing managers held a conference on its report on homelessness (Institute of Housing 1988). At this time, the Department was – at the Government's request – investigating options for repealing the Homeless Persons' Act, or further restricting its application, but eventually the Government decided to accept the *status quo*.

In so far as the policies of the British Governments and the US Federal Governments were defended, the argument was that homelessness was voluntary. The argument was similar to that underlying the Vagrancy Acts. 'A vagrant was thought to be one who was able to work for his maintenance, but preferred to live idly, often as a beggar' (Encyclopedia Britannica). This deterrent policy was ineffective in reducing homelessness, since it failed to recognize its causes. In France, Germany and the Netherlands, homelessness seems to have enjoyed more official sympathy. In France, the 1989 Act makes a clear commitment to reintegrate the homeless into society.

Howenstine points out that there have been positive changes in recent years in the USA. The appointment of Mr Jack Kemp has ushered in a more open and honest administration at HUD – even though his campaign to 'empower' the inhabitants of the ghettos has received little support from the Republican Party, the President, or Congress. Federal assistance for shelters has been provided under the 1987 Homeless Assistance Act, and there have been local initiatives for 'eviction avoidance'. HUD supports these initiatives, but most of the expense falls on the cities. Many cities, however, face severe financial difficulties, and are now having to close down facilities for the homeless. The immediate cause of this crisis in local finance is the recession, but the underlying cause is the dependence on a notoriously inelastic source of finance – the property tax. Thus the attack on homelessness and the other urban ills of the USA will probably be hamstrung as long as cities (and even States) lack a tax base which corresponds to their responsibilities.

Effective programmes against homelessness in the USA and UK – and the other countries studied – would have to include a range of policies for both prevention and cure, going beyond housing policy as such. They would have to address underlying issues: high unemployment, the growing polarization of incomes (reflecting in part the lack of skills in the underclass), family breakdown, crime, and the tendency for (public and private) urban renewal to reduce the supply of low-rent housing. These are intractable issues, going to the heart of our economic system. However, many small steps, costing relatively little money, could be taken to help a lot of homeless people in the 'economic' and 'situational' categories to break out of a vicious circle. Counselling, training (where necessary), and the judicious one-off payments (e.g. for deposits on rented accommodation) could help many people to avoid becoming homeless. Since – even with preventive policies of this sort – some people will still become homeless, local authorities need to have some subsidized housing available for emergency cases.

### Income support or housing subsidies?

The contributors' recommendations on housing affordability relate mainly to housing allowances, social housing, and tax concessions. There is, however, another type of policy which deserves mention. The radical view that a comprehensive system of income support would eliminate the need for any kind of housing subsidies was put forward thirty years ago by the economist Milton Friedman, who was the best known (although not the first) advocate of a 'negative income tax' (Friedman 1960). Friedman proposed that households under the income tax threshold should receive a payment which rose as income fell. It would then be possible, he argued, to abolish all other welfare payments in cash or kind, including social housing. The idea has since been advocated under names such as 'basic income' or 'guaranteed minimum income'. However, no country has introduced a scheme on these lines, except for that introduced in 1989 in France, while expenditure on housing allowances has mushroomed.

The main argument for a 'basic income' scheme is the libertarian one that an untied payment allows full scope for consumer preferences. If a household chooses to spend little on housing and more on other things, why should 'nanny state' interfere? The converse argument, which seems to find more acceptance, is that a housing allowance is spent on something worthy of public support, whereas an untied payment could be spent on drink. However, the 'basic income' scheme is also open to objection on cost grounds. Even economists who are sympathetic to the idea have in recent years accepted the criticism that, if the basic income were set at the level of a 'living wage', high rates of income tax, with consequent (possible) disincentive effects, would be necessary. It seems doubtful whether a basic income scheme could satisfactorily replace housing subsidies, although there are still good arguments for introducing it at a low level, as the French have done. In the meantime, public authorities framing policies to deal with housing affordability have to decide what mixture of social housing, housing allowances and tax concessions to use.

In the 1970s and 1980s, the argument was frequently advanced for 'subsidizing people rather than houses', and the housing allowance has indeed several advantages as compared with social housing. It gives assistance to everyone on low incomes, not only tenants of social housing; it is, according to some studies, considerably cheaper than social housing. For these reasons, all the contributors accept that it has a major role to play, but they also point out why it cannot be a complete solution. It does not provide affordable housing directly, but merely gives low-income households the means to compete more effectively in the market place. The extent to which this will help them will depend on the extent to which there is a well-functioning private rental market. In practice, private rental markets often suffer from a variety of failings. There can be discrimination against various minority groups. When the private market has fallen into disuse and disrepute (as in Britain) it may need a long convalescence. Moreover, there are often substantial differences in house prices and rents between regions or districts. Thus a (uniform) national housing allowance scheme can still leave low-income people virtually excluded from certain districts, and facing severe difficulties in others. Social housing, on the other hand, can be provided in the districts when there is a shortage of affordable housing, and will benefit tenants directly and immediately.

Finally, there are the needs of what the Victorians called 'the incompetent poor'. There is a significant minority of people with mental disabilities who do not have the skills needed for filling in housing allowance forms, and finding accommodation, even if they did not face discrimination. In the past, many were out of sight in 'asylums', while others found a niche in various unskilled jobs offering accommodation. Now that these havens have largely ceased to exist, the need for social housing (allied to social care) has increased. This point is conceded by most critics of social housing, who would allow 'special needs' housing. However, persons with 'special needs' are usually defined as those with severe physical or mental disabilities. There is a much wider group of society's casualties.

Social housing has often been criticized on grounds of cost, but housing allowance systems can also incur substantial administrative costs. One study gives the following figures of the administrative cost as a percentage of the amount paid; Sweden, 5 per cent; Germany, 8.6 per cent; France, 12–15 per cent; UK, 17 per cent; USA, 22 per cent (Howenstine 1986, p. 68). In the UK, the system has been regularly thrown into chaos as a result of arbitrary changes by central government, introduced without consultation with local authorities (Kemp 1984). A more general defect of the housing allowance is that the take-up rate is often low. The contributors of the chapters on Britain, West Germany and the USA mention rates of around 50 per cent although, in the Netherlands, among people with low incomes, the take-up is around 75 per cent. Thus although the British and US systems would be cheaper and more effective if they could move towards the German, Dutch or Swedish levels of competence, the housing allowance system can still miss some households in the greatest need.

For all these reasons, there is a strong case for maintaining a social housing sector, but the most satisfactory size and type can be established only by experience. In the USA, the UK and West Germany, governments in the 1980s experimented with cutting back social housing and replacing it with housing allowances. Conditions varied from country to country, but Howenstine, Malpass, and Ulbrich and Wullkopf all agree (with different levels of emphasis) that the experiments were not wholly successful, even if the privatization of some social housing units was justified. In France and the Netherlands, there was a similar move towards housing allowances, although the stock of social housing was maintained. All the contributors to this book conclude that there is a continuing need for a social housing sector – even if its character and organization are different from what they were in the past.

### The organization of social housing

The challenge to social housing has reduced the size of the sector (in the UK and West Germany), but it has also stimulated a response. New types of 'customer-based' social housing are being developed. In the USA and the UK, in particular, there have been moves to 'empower' tenants, ranging from increased legal rights to involvement in management. This trend to 'empowerment', which represents one of the more positive developments in the housing scene in the USA and the UK in the 1980s, seems set to be a major concern of social housing in all the countries studied.

As part of this trend, the 1990 HOPE programme in the USA is, on the face of it, an admirable programme embodying a variety of types of low-income housing. It accepts the need for affordable rented as well as owner-occupied housing; it emphasizes the role of community organizations and the need for them to match Federal funds; it requires that any housing sold should be replaced. Howenstine explains why the outcome may well fall short of the published ideals. Nevertheless, the programme, and the growth of community-based housing associations, suggests that an encouraging new chapter in US housing policy has begun.

In Britain, the legislation allowing tenants to choose their own landlord, and the emphasis on the quantitative assessment of service provision (even if motivated by a hostility to council housing) have encouraged more decentralized and effective management. Some councils practised decentralized management from the beginning but, in many of the larger cities, decision-making was too centralized, resulting in poor maintenance and indifference to tenants' wishes. The management of council housing is now being transformed. In some British cities (e.g. Glasgow) 'management cooperatives' have been formed to manage, although not own, council housing. In a few cases (e.g. Liverpool) tenant-owned cooperatives have been formed. Some have worked very well; others have not; the presence of a stable body of tenants is crucial.

In France, recent housing developments (ZACs) have been smaller in scale, and better managed than previously, and with a mix of tenure and price. There have also been comprehensive programmes for neighbourhood development (*Development Social des Quartiers*) involving structural improvements, but also educational, social and employment programmes. In West Germany, *Neue Heimat* was successfully broken up into smaller units.

### Councils or housing associations?

In Britain, social housing is undergoing a greater upheaval than in the continental European countries, partly because of the previous concentration on council housing. Social housing is generally defined as housing which is financially assisted by the state, so that rents can be held at below market levels, to assist low-income families. However, although subsidies are provided by the state, the housing can be delivered by a variety of organizations. In the UK (and the USA) there was originally an assumption that social housing should be administered directly by a local authority. The UK has experienced a swing from a virtual local authority monopoly of social housing between the 1920s and 1970s to a planned elimination of local authority housing in the 1980s. Such large policy swings have not occurred in the continental European countries.

The determination of the post-1979 Conservative Governments to eliminate council housing was not based on any inherent enthusiasm for housing associations. However, housing associations were chosen as the destination for local authority housing which could not be sold to tenants. As Malpass points out, the 'MacClennan/Williams' studies on the relative efficiency of local authorities and housing associations were instituted by the Government in the hope that they would show up the inefficiency of local authorities. The reports reached a different conclusion. The housing associations, which were much smaller and with much higher ratios of staff to housing, produced better services, at a substantially higher cost per dwelling. There were substantial differences between local authorities, indicating scope for the poorer to catch up with the better, but the main conclusion was that 'you get what you pay for'.

Whether tenants should have a high level of service and high rents, or a low level and low rents (provided that the basic repairs needed to maintain the buildings' upkeep are sustained) is not capable of a simple answer. If one believes that tenants' interests are paramount, it is hard to see why tenants should not be asked to vote on alternative 'packages'. This system is common in Scandinavia but, in the countries of our study, the original 'top down' management systems are still in the process of adaptation to a 'client-orientated' approach.

One argument put forward for removing housing from council control is that no other European country gives local authorities direct responsibility for social housing. This is true – up to a point. However, neither France, West Germany nor the Netherlands relies solely on independent housing associations, in the way currently envisaged for the UK. In West Germany, the larger cities are majority shareholders in housing associations; the management of the association may be 'at arm's length' but the city lays down general policy. Moreover, *Neue Heimat* illustrated the problems that can arise in the management of large, independent housing associations, and there has been at least one similar case in the UK (Hallett 1991).

France relies on quasi-governmental regional organizations, partly because many of the smaller of France's 3,500 communes would be too small for the effective provision of social housing, but the communes play a role in allocating housing. In the Netherlands, where the municipalities had also been the main providers of social housing in the post-war period, it was decided in 1965 that most social housing would in future be undertaken by housing associations. Housing associations have gradually become dominant, but the municipalities have not been compelled to divest themselves of their housing. Voluntary transfers have taken place in a few cases. The experience of France, Germany and the Netherlands certainly shows that there are workable alternatives to the original British system, under which local authorities provided social housing directly, and enjoyed a virtual monopoly. The continental European experience cannot, however, be used to justify the complete exclusion of local authorities from the provision of social housing, or at least some say in its allocation.

### The emasculation of local authorities in Britain

Whereas the UK is, in some ways, converging with the Continent of Europe in adopting a more pluralistic organization of social housing, in other ways it is diverging from it. The UK was once unusual in giving local authorities a virtual monopoly of social housing (and, to a large extent, even of rented housing in general); it is now unusual in the extent to which local authorities have been emasculated. In the 1980s, there was an international trend to decentralize governmental functions, including those concerned with housing. In the USA, President Reagan supposedly presided over a New Federalism, although it consisted mainly of halving Federal aid to the cities without giving them alternative sources of revenue (Hambleton 1988). West Germany maintained its federal system, and devolved housing functions away from the Federal Government. France undertook a major administrative devolution. In the Netherlands, the 1989 policy proposals for 'Housing in the 1990s' advocate administrative decentralization. In the former GDR, the *Laender* and communes have recovered their powers.

In the UK, by contrast, there has been a steady centralization since 1979, both in general local–central government relations and in the finance of council housing. Under the arrangements introduced in 1991 to replace the poll tax, the central Government each year sets the level of expenditure for every local authority. Under 10 per cent of local expenditure will be raised from a local 'council tax' on housing, but the rate for each authority will be 'capped'. Thus local authorities have in effect no power to vary local expenditure in response to local needs. Similarly, one study of council housing concludes that 'In retrospect, the 1980s system may be seen simply as a staging post on the route from a two-tier system to a pure agency model' (Kleinman 1989).

The current British system of central–local government relations has no parallel in the other countries studied, except for the system that prevailed in the German Democratic Republic. A system in which local authorities are merely agents of the central government may be adequate for a programme of large-scale state-financed housebuilding or for a policy of *laissez faire*; it is not well suited to tackling varied local housing problems in a variety of ways. Some Government spokesmen have talked of an 'enabling role' for local authorities in housing i.e. coordinating the work of other organizations. Even such a limited role, however, requires finance and administration, which many local authorities would now find it difficult to provide.

Moreover, in 1991 the Government announced another major reorganization of local government; it plans to reverse the reorganization carried out in the 1970s by abolishing all upper-tier 'county' authorities. The current division of responsibilities between the two tiers is – it is generally accepted – often unsatisfactory, but a radical reorganization would be very disruptive. The previous reorganization caused chaos in local administration for most of the 1970s, and the new reorganization (if implemented in the sweeping way proposed) can be expected to do the same in the 1990s. It is even doubtful whether a satisfactory structure will emerge at the end. All other European countries have found that 'regions' or 'counties', or associations of authorities, are needed for functions such as town planning and transport planning which affect large areas – and have an important, although indirect, bearing on housing affordability.

A few lone voices call for a revival of local government by giving local authorities independent sources of revenue, and creating a new partnership with central government (Carter and John 1992). There is, however, no prospect of such a reversal of current trends in the 1990s. British local government is likely to remain weak, and its weaknesses (supporters of local democracy argue) would undermine even the best conceived of national policies for homelessness and low-income housing.

## GUIDELINES ON HOUSING AFFORDABILITY

The contributors to this book have put forward many moderate and reasonable proposals for improving housing affordability for low-income households. The principles underlying their proposals can perhaps be summarized as:

1 Housing policies should be based on a consistent treatment of households, with the greatest help going to those in greatest need. (A social criterion).
2 Landlords' receipts (net, after taxes and subsidies) should be sufficient to maintain the housing in good condition. (An investment criterion).
3 The allocation of public expenditure between different programmes should be based on their relative cost-effectiveness. (An economic criterion).

It is only too clear from the contributors' accounts that housing policies have not always been based on principles of this sort; hidden political agenda and interest group pressures have often been far more important. However, in the hope that rational argument has some influence in the long run, let us review public policy on 'housing affordability' in the light of these three principles.

*Principle 1* raises the question of how one measures 'need'. What in fact do we mean by 'affordable'? This is a subjective and complex subject (discussed in the chapters by Howenstine and Malpass). The current US practice of setting a uniform 30 per cent limit is criticized by Howenstine for bearing particularly hard on large, poor families. Howenstine mentions the 'market basket' method of calculating how much a household can afford: i.e. to estimate the minimum expenditure on items other than housing, and to deduct this amount from disposable income. He points out that, on this basis, many households in the lowest income group in the USA could not afford to pay anything like 30 per cent of their income for housing. If they in fact pay this percentage, it is at the cost of deprivation in other ways.

The use of a single percentage for all households is clearly unsatisfactory; any 'standard' percentage should be reduced for large families, and for very poor families – as is implicitly or explicitly accepted in most European housing allowance systems. Some systems take the form of a complex table, with variations according to income and family size, as in the German and Dutch housing allowances. A different arrangement (used in the UK and Germany for welfare payments to the poorest households) is for all housing costs to be paid by the state. In France, the rule is that a minimum sum, often 200 FFR a month, has to be paid, as it is considered important that everyone knows that housing has a cost.

In spite of the defects of the current US system, the practice of laying down a 'standard' percentage figure (from which deductions can be made for special family circumstances) has many virtues; it provides a yardstick for public discussion, and for rent-setting in social housing. The 'traditional' 25 per cent (used as a guideline in France) still seems a reasonable figure. (It should be added that the current '30 per cent' in the USA includes local taxes, heating and lighting; eliminating these items would bring the level below 25 per cent.)

The principle of giving most help to the needy is often breached in connection with mortgage tax relief (see below) but also in some arrangements for social housing. In the UK, council rents (by an indirect but very effective procedure) are now set by the central government for every local authority, and are being raised towards market levels. The end result would be that council housing would cease to be social housing. The Government's objective, however, (which became increasingly explicit in the various Acts passed in the 1980s) has been to force councils to sell their housing to tenants, or transfer it to housing associations. This may, or may not, be a reasonable objective, but the financial regime which has been introduced to achieve it functions (rather like the Poll Tax) by bearing hardest on the poorest council tenants.

The current financial regime for British housing associations is completely different from that for council housing. It involves a grant at a fixed percentage of the permitted cost of a project, combined with private finance; associations are then free to set their own rents. This system has the virtue of giving associations every incentive to minimize costs and (in theory although not always in practice) considerable freedom of action. It also has defects. The rents set by the associations are supposed to be 'affordable', but no definition of this term has been given. Rents have risen sharply, and many tenants are paying over 30 per cent of their income. Further consequences are that the increased risk associated with rehabilitation (following the ending of any retrospective grants) has caused associations to virtually withdraw from this very necessary activity.

*Principle 2 (Investment).* The point was well put by a housing engineer in East Berlin. 'If you don't spend 2 per cent of a building's value annually on maintenance and improvements, in thirty years you will have a ruin.' One can argue about the percentage – which other experts put at around 1 per cent – but the basic point is unarguable. Examples of inadequate maintenance outside the former GDR can be found in rent-controlled private housing in the UK and (although dating from an earlier period) in France. In Britain, the Audit Commission has criticized councils for not setting rents at a level high enough to make regular maintenance possible. This criticism may not be unjustified but, as Malpass points out, there are two other considerations: (a) if tenants cannot afford to pay the necessary rents, it could equally well be argued that the subsidy was too low; (b) the central Government, by its politically motivated limitations on local government housing expenditure, has itself been an 'engine of disrepair'.

*Principle 3* is often breached because 'welfare' and 'housing' expenditure, although intimately related, are often handled independently. For example, in the UK, many homeless households have been put into unsuitable private 'bed and breakfast' accommodation, (costing around £15,000 per head p.a.) when building social housing, taking long leases on private housing, or even using mobile homes, would both give an incomparably better environment (especially for children) and be much cheaper. This anomalous situation has arisen for various reasons: because different government departments, with separate budgets, are involved; because of a feeling that homeless people are 'scroungers' who should be punished; and because of the Government's campaign against council housing.

Moreover, directly providing households with housing is not the only, and not always the most appropriate, means of resolving their housing problems. Current 'housing problems' are very diverse, with big differences between districts and between households. One household may need only debt counselling; another financial assistance; another specialized housing and social care. This suggests that powers to assist households facing serious housing problems should be decentralized, and that local agencies should be able to provide whatever type of assistance will meet individual household's housing needs most cheaply. This genuine type of 'enabling role' is gaining ground in many countries, and is most explicitly expounded in the French 1989 Act.

## THE ROLE OF OWNER-OCCUPATION IN LOW-INCOME HOUSING

A recurrent theme in the national chapters is the preferential treatment given by governments to owner-occupation, through tax concessions and, in some countries, the sale of social housing to tenants on concessionary terms. Peter Malpass is particularly critical of the Conservative Governments' 'tenure policy' – meaning in this case a preoccupation with expanding owner-occupation. Malpass argues, and most students of British housing would agree with him (even if they may not all agree with his dismissal of private rented housing), that the emphasis on owner-occupation has reduced the supply of affordable rented housing, and sometimes induced households to embark on home-ownership which they cannot finance.

Criticisms of the tax concessions for owner-occupation are not confined to Britain. In Germany and the USA (and, almost certainly, Britain, although the figures have not been published), the tax concessions for owner-occupation give more assistance to the rich than other housing subsidies give to the poor. As Howenstine puts it, describing the particularly striking situation in the USA, the subsidies are going to the wrong people!

Is it possible, in the light of national experiences, to reach any general conclusions about policy towards owner-occupation (and private tenancy and the non-profit sector) in a housing strategy designed to meet Principle 1, i.e. to help the poor rather than the rich? At this point, value judgements (not to mention political tactics) rear their heads. Conservatives have often taken the view that owner-occupation is 'a good thing', and so should be encouraged; some socialists have argued that owner-occupation is a 'bad thing', and private tenancy a very bad thing. However, as rented housing shrinks and owner-occupation increases, some socialists are becoming converted to the need for private rented housing and owner-occupation, and some conservatives are beginning to question the wisdom of pushing owner-occupation too far. The idea of 'tenure neutrality' seems to be gaining ground. Tenure neutrality means that the state should not favour one tenure over another, by setting discriminatory financial conditions. However, within a broadly 'tenure neutral' policy, assistance for owner-occupation could be targeted on households with low incomes (while maintaining the supply of affordable rented housing).

There are three main ways in which owner-occupation is assisted in the countries studied.

1 The tax-deductibility of mortgage interest.
2 Subsidized loans, or grants, for low-income households.
3 Subsidized savings-for-purchase schemes.

A fourth variant, which has been proposed in Britain, is 'mortgage benefit'.

### Tax-deductibility of mortgage interest

Whether tax relief on mortgage interest is a 'subsidy' has been the subject of considerable discussion, but the issue is basically straightforward. Many countries originally had a system in which income tax on 'imputed rent' was balanced by tax deductibility of mortgage interest. Such a system will be 'tenure neutral' – provided that the imputed rent is based on current values. If the tax on imputed rent is dropped, but the tax deductibility of mortgage interest retained, the tax deductibility favours owner-occupation as against tenancy, and can be considered a 'subsidy' (or a 'tax expenditure'). This has tended to happen, with only the Netherlands retaining a tax on imputed rent. Disallowing tax-deductibility would remove the subsidy for borrowers, but not for people with capital, who do not need to borrow. Thus taxing imputed rent is theoretically preferable to ending tax deductibility, although most certainly less feasible politically.

### Tax relief on loans for owner-occupation

*USA*: Interest on mortgage debt of up to $1 million can be set against Income Tax.

*UK*: Interest on mortgage debt of up to £30,000 per household can be set against Income Tax, at (since 1991) only the basic rate.

*West Germany*: (Since 1988) A sum equal to 5 per cent of the value of a house, up to DM 300,000, can be deducted for eight years, 'once in a lifetime'.

*France*: Interest deductible from taxable income up to a limit of FFR 3,000 p.a. for five years.

*Netherlands*: Interest on mortgage debt fully deductible from taxable income. Tax on 'imputed rent'.

All the countries (except the Netherlands, which taxes 'imputed rent') set some limit for tax-deductibility. However, the USA, with its maximum of a million dollar mortgage (and a further million on a second home) goes way beyond anything available in Europe. France is at the other extreme, with an almost negligible tax deduction, and an emphasis on subsidized, targeted loans. In Britain, the real value of the £30,000 limit has been reduced over the years by rising house prices. But even the current British system still leaves most of the benefits going to higher income households which would have been owner-occupiers anyway. The second report of the 'Inquiry into British Housing' chaired by the Duke of Edinburgh, repeated the proposal in its first report that tax deductibility should be gradually phased out, and replaced by a scheme for giving assistance to low-income households with their housing costs, whether they are tenants or owner-occupiers (Edinburgh 1985, 1990).

Germany changed its tax treatment of owner-occupation in 1988. Previously, a tax on imputed rent (although at an unrealistically low level) had offset tax deductibility. The Government then abolished the tax on imputed rent while retaining tax deductibility in a new form. The new scheme has the virtue of being 'once in a life-time' and limited to eight years, but it is poorly targeted. It has now been extended to the five new *Laender*, where it is particularly inappropriate because of the low level of incomes.

A 'tenure neutral' policy can be achieved either by giving no concessions to any tenure, or by giving comparable concessions to all tenures. Tax concessions to private landlords, or subsidies to social housing, are not a 'distortion of competition' if they merely balance tax concessions to owner-occupation. One 'distortion' can offset another ('the theory of the second best'). In practice, it is difficult to determine precisely when even-handedness between tenures prevails, but easy to tell when there is a marked lack of it. The UK has discriminated against private tenancy by (until very recently) imposing punitive rent controls, and by not providing any counterpart to mortgage interest tax deductibility. West Germany, France and the Netherlands (and to some extent the USA) have maintained a better balance, through less restrictive rent policies and tax breaks for private landlords. Similarly, statisticians found it difficult in the 1970s to decide whether owner-occupation or council housing in the UK were more heavily subsidized (which probably suggests that there was not much in it either way). However, with the sharp cut-back in subsidies to council housing in the 1980s, the balance shifted decisively in favour of owner-occupation. In Germany, France and the Netherlands a better balance between owner-occupation and social housing has been maintained.

It is also possible to modify the arrangements for tax-deductibility so as to achieve more equity between owner-occupiers. A deduction from taxable income is inherently regressive, given a progressive income tax. A deduction of taxable income of £1,000, for example, is worth £500 for someone paying 50 per cent marginal tax; £250 for someone paying 25 per cent; and zero to a non-taxpayer. An alternative proposal is that the deduction should consist of a 'tax credit' (i.e. an amount of tax) rather than a deduction from taxable income; this would ensure that all taxpayers were treated equally. Howenstine advocates this system for the USA. The decision by the British Government in 1991, that relief will be at only the basic rate of tax, is in effect the same thing.

A change of this type does not benefit non-taxpayers, but arrangements have been in force in Britain since 1968 which give non-taxpayers a cash payment instead of a tax relief. This 'option' scheme was based on a proposal by two economists (Merrett and Sykes 1965); it must rank as one of the few proposals by economists which have both been implemented and have proved beneficial.

### Subsidized loans or grants

These play an important role in the Netherlands, Germany and France. In Germany, the 'Second subsidy way' provides a subsidy to lower income households. In France, the PAP (*Prêt pour l'Accession a la Proprieté*) is a subsidized loan for lower income households which can be used for purchasing new or old dwellings, or for refurbishment. In principle, these schemes are better targeted than the tax-deductibility of mortgage interest.

### Contractual saving-for-purchase schemes

Some countries provide subsidies to people who enter into contractual savings schemes for house purchase. The most developed scheme is the German 'Building savings premium'. Such schemes concentrate assistance on households which are on the threshold of owner-occupation, and ensure that house-buyers have some capital of their own – and so are less likely to find themselves over-borrowed. In other words, these schemes both provide a more targeted form of assistance than tax deductibility and also encourage financial prudence. However, instead of giving priority to such schemes, Governments tended to downplay them in the 1980s. In Germany, expenditure on savings premiums fell in relation to tax-deductions, and in Britain a 'Homeloan' savings scheme, introduced in 1978, was abandoned in 1986. This change of emphasis may have been connected with the liberalization of housing finance (especially in the USA and UK) and the 'get rich quick' philosophy which accompanied it. In the painful aftermath, politicians have started preaching the virtues of saving. Perhaps this belated conversion to 'Victorian values' will encourage a switch in emphasis from tax deductibility to savings schemes. A step in this direction is the provision in the HOPE programme in the USA to allow withdrawals from tax-free IRAs (Individual Retirement Accounts) for a down-payment on a home.

**Mortgage benefit**

Another way of assisting low-income owner-occupiers, which has been advocated in Britain but not yet introduced anywhere, has been called 'mortgage benefit'. The basic idea, as advocated by the commission chaired by the Duke of Edinburgh (Edinburgh 1985, 1990) and by the Liberal Democrats, is to (gradually) replace mortgage interest relief with a scheme which gives assistance to low-income households, whether they are tenants or owner-occupiers. A detailed examination of possible schemes has recently been made (Webb and Wilcox 1992).

The proposal is that low-income owner-occupiers should be eligible for assistance with their interest payments, on a sliding scale. The amount of assistance would vary with the household's income and the proportion of income paid on mortgage interest – in a comparable way to 'housing benefit' (which applies only to tenants). The argument is that low-income tenants are assisted by housing benefit, while unemployed households who are eligible for 'income support' have their mortgage commitments paid by the social security authorities. There is, however, no comparable assistance for households who have suffered a fall in income, e.g. as a result of short-time working, or the unemployment of one partner. Many households in Britain are now in this situation, and some are threatened with repossession. Filling the gap in assistance (using revenue saved from a reduction of mortgage interest relief, e.g. from 25 to 20 per cent) would, it is argued, be justified on grounds of equity and efficiency. It would contribute to resolving the current repossession crisis and would also remove the disincentive effect of the present system, under which an unemployed household can become worse off if one partner takes a job.

This proposal differs from the arrangements for assistance to low-income households which exist in France, Germany and the Netherlands. The schemes in these countries base the level of assistance on the household's income when the mortgage is taken out. This was satisfactory in the period from the 1950s to the 1970s, when employment was high and stable, but there is now much less security of employment and income in all the countries studied. There is thus a strong case for reducing 'across the board' support for owner-occupiers, and introducing the 'safety net' of a scheme which varies the amount of assistance according to the burden of all housing costs in relation to income.

**Mixed tenures**

This discussion has, so far, stressed the defects of what Malpass calls 'tenure policy'. But this is not to deny that there is a widespread desire among households to build up ownership in housing, which should be facilitated. It should be possible for households to start on the path to ownership without running the risks demonstrated by the current wave of repossessions in the UK. One way is through the development of mixed tenures, in which the occupier is part-owner and part-tenant.

This idea has recently been taken up in Britain, and some experiments in 'shared ownership' and 'flexible tenure' have been carried out by housing associations. There are many potential advantages for lower- or middle-income households in arrangements which would provide more flexibility – in both directions – between ownership and tenancy. Young households could build up equity, without taking the plunge into full ownership; elderly households on low incomes could withdraw equity to maintain their current expenditure; owners facing repayment problems could become tenants. There is scope for the ingenuity displayed in the development of 'flexible' financial products to be extended to tenure arrangements (National Federation of Housing Associations 1990). The 'rents into mortgages' scheme for council tenants announced by the British Government in May 1992, is *not* of this type, for it allows movement in only one direction.

## THE ROLE OF PRIVATE TENANCY IN LOW-INCOME HOUSING

Does the private rented sector have a role to play in the provision of low-income housing? In all the countries studied except the UK, there is a viable, even if declining, private rented sector, and in West Germany and the USA, social housing has been provided through private landlords. In the UK, by contrast, it has been a matter of controversy whether private rented housing should, or could, continue to exist. Its elimination has been an implicit or explicit aim of Labour Governments. Mr Harold Wilson, the Prime Minister, in 1969 expressed the view held by the Labour Party since its foundation: 'We do not consider the provision of rented accommodation to be an appropriate activity for private enterprise'. This view was shared by members of the dominant 'Titmuss' school of academic housing studies in the 1960s and 1970s, who argued (somewhat inconsistently) that rent controls had no effect on the supply of private rented housing, and that private tenancy was an objectionable system, whose demise was to be welcomed (Hallett 1977). This was the background to more restrictive and longer lasting rent controls than in any of the other countries, which Conservative Governments, before 1988, were unwilling or unable to alter. The sector has declined more sharply than in the other countries; it currently accounts for only 7 per cent of the housing stock, and the percentage is still falling.

Attitudes to private renting, however, are changing. In 1991 the Labour Party published a discussion paper which proposed a two-tier market; above a certain level, rents would be uncontrolled, but tenants would not be eligible for the housing allowance (housing benefit); below it, rents would be subject to a mild 'cost-covering' control, but landlords would be eligible for grants, and tenants for housing benefit. Such a scheme was not only a reversal of previous policy; it arguably offered a better environment for the development of 'affordable' private rented housing than that introduced by the Conservative Government. There was, however, some back-tracking in the subsequent Party document 'Welcome Home'.

New tenancies were decontrolled in 1988, (which, in the short term, led to some households becoming homeless, whatever the long-term benefits). However, the opportunity was missed of transforming 'Fair Rents' (which were economically meaningless) into a form of rent arbitration, comparable to the German and Dutch 'comparative rents', and the French equivalent, or 'fair market value' in the USA. Such a scheme would have benefited reputable landlords as well as tenants, and helped to ensure broad political support for the ending of 'rent control'.

In spite of Conservative decontrol, and Labour revisionism, a revival of the private rented sector in Britain is far from guaranteed. Countries with a viable private rented sector, such as Germany and France, have had a reasonably consensual policy for many years, which has underpinned investor confidence. Moreover, landlords enjoy tax breaks, comparable to those given to owner-occupiers, and the backing of financial institutions. Even in the Netherlands, with its massive intervention in the housing market, financial institutions have moved into rented housing, and now account for about 10 per cent of the stock. These situations do not yet exist in Britain. The only fiscal support for private tenancy introduced by the Conservative Governments, the Business Expansion Scheme, was devised for quite different purposes, and is unsuited for the encouragement of long-term housing investment. It gives tax relief for investors in rented housing for six years, after which all controls are lifted. Investigations suggest that many investors will seek to liquidate their investments when the schemes expire.

Since 1990, advertisements of houses and flats to let have begun to appear, but they are mostly by owners who have been unable to sell them, and will not continue letting them when the market revives. Nor is there the institutional involvement found in Germany and the Netherlands. Before 1939, insurance companies invested in private rented housing, but no one in business today remembers those days. Nor – after the traumas of the past half-century – are most individuals keen to invest in rented housing. A study of housing inheritance has shown that inherited rented housing is nearly always sold, so that there is a continuing haemorrhage of the sector (Hamnett, Harmer and Williams 1991).

The British problem – of reintroducing a species which has virtually died out – is more difficult than the problem in the other countries, of maintaining a stable or slowly falling population. The main need is for tax breaks to offset the concessions to owner-occupation and the various official 'tax havens' which have been introduced in recent years; confidence that the financial arrangements will not be subject to short-term political changes; and an improvement in the status of the private landlord.

### The shortage of cheap rented housing

In countries other than Britain, private tenancy is better established, but there is widespread concern over a decline in 'affordable' private rented housing. In West Germany, private tenancy has tended to go up-market, with de luxe modernization and much higher rents. This trend is partly the result of the country's geographical situation, at the centre of the 'hot banana' of economic growth curving from London down to Lombardy. It is, Ulbrich and Wullkopf suggest, also the result of excessive public support for 'urban renewal' as against new building.

In the USA, the private rented sector plays a more crucial role than in Europe, because of the extremely small 'public housing' sector. Most public expenditure has taken the form of subsidies to private landlords. Howenstine concludes that 'Tax concessions have been an effective tool to induce individuals and corporations to invest in low-income housing'. He draws attention, however, to the sharp decline in the stock of low-income housing, resulting from demolition or conversion into owner-occupation (condominiums). Moreover, many of the publicly-supported schemes are, as in Germany, due to come to an end in the 1990s.

The two basic questions are: (a) can the commercial sector make any contribution to low-income housing; (b) should social housing schemes be extended to private landlords? On (a) Howenstine points out that, although unsubsidized private landlords cannot, in the nature of things, be expected to provide much housing for very low-income households, they can often cater for households with slightly higher incomes. By drawing off such households, unsubsidized private tenancy can free social housing for lower income households.

The second question is whether privately owned housing should be used for social housing schemes. There are advantages and disadvantages. In post-war West Germany, the extension of social housing to private landlords utilized resources of capital and management which would not otherwise have been available to social housing. On the other hand, German experience shows that private tenancy cannot be relied on to provide social housing for the long term. If the need for social housing is regarded as more than temporary (which, after the experience of the 1980s seems the wisest assumption to make) then it would be better to rely primarily on public or non-profit organizations – if this is politically possible. If it is not, it makes sense to combine limited public investment with fiscal incentives to private landlords.

## THE NON-PROFIT HOUSING ENTERPRISE

The 1980s have seen divergent trends in non-profit housing in the countries studied. In the USA, new 'community' non-profit housing associations are rising like a phoenix from the ashes of 'public housing'. In Britain, the movement was revived in the 1970s, after half a century of virtual council housing monopoly; it acquired a new importance as councils began to be forced to dispose of their housing. In Germany, *Neue Heimat* collapsed, and its fall contributed to the abolition of the legal status of 'non-profit housing enterprises' (apart from cooperatives). In France and the Netherlands – where the enterprises have always had a quasi-official character, there has been more stability. In both countries, the tendency is for the enterprises to become less dependent on local or central government, and to assume the responsibility for constructing, renewing and managing housing in a more commercial spirit.

The rationale of the non-profit housing movement has long been debated. Is it an 'alternative' economic system, or a useful part of the 'capitalist' system, or a fraud? The most explicit theoretical debate has been in West Germany (Hallett 1977 p. 62). In the 1970s, the Left criticized the movement, on the grounds that it was a capitalist wolf in sheep's clothing. Critics from the Right, on the other hand, argued that non-profit enterprises at best did what profit-making enterprises did, but less efficiently. 'Revisionists' within the movement argued that, although it was not – as some of the founders had believed – an 'alternative' system, there were ways in which it could serve a special 'social' purpose. First, it could protect tenants from the effects of inflation, by giving them the benefit of rising house prices which would otherwise go to a private landlord. Second, it could sometimes give more emphasis to the needs of the disadvantaged than a purely commercial enterprise. Third, it could act as an agent for public spending (e.g. on social housing or urban renewal) while being free of the possible conflicts of interest of a profit-making firm, and the 'bureaucracy' of the civil service. The abolitionists carried the day in Germany – although most of the 'formerly non-profit enterprises' remain, and are likely to continue to operate with a 'social' slant. In France and the Netherlands, where it was felt that local authorities should not control all social housing, non-profit housing agencies and companies have a long, and generally impressive record. They are increasingly coming to play this role in Britain and the USA. Germany's experience with *Neue Heimat*, however, suggests that there are dangers when non-profit enterprises become too big. The directors lost an overview of the financial situation and the regulators failed to detect what was happening, probably because they were 'captured' by such a large organization. There would seem to be lessons for Britain, where the housing associations are under pressure to become larger.

One of the most significant recent commentaries on the non-profit movement has come from a commentator on big business. Peter Drucker (1990) foresees the non-profit movement becoming more important and more 'businesslike', while profit-making businesses acquire some of the characteristics of non-profit enterprises. He regards non-profit enterprises as businesses which are concerned to identify their consumers and give them what they want, even though this does not happen through the profit-maximizing mechanism. This is already the ethos of the non-profit housing agencies and companies in France and the Netherlands and, increasingly, the UK.

Drucker also believes that, for the highly professionalized employees who will increasingly set the tone of 'ordinary' business, the achievement of ever higher profits or incomes will not be sufficiently satisfying, and that they will offer some of their time to charitable non-profit organizations. Written at the end of a decade which – in the 'Anglo-Saxon' countries – has become a by-word for the ruthless and even illegal pursuit of personal gain, that might seem a vain hope. However, a less reported side of the USA in the 1980s was 'voluntarism' – the growth of groups of concerned citizens, who set about improving conditions in their communities. These groups sometimes developed into substantial non-profit organizations such as the Community Development Corporations, which organize housing and housing-related initiatives (OECD 1992). If this trend crosses the Atlantic, the West German decision to abolish the status of non-profit housing enterprises will be seen to have been a mistake, based on a blinkered economic theory of society.

## THE COUNTRIES COMPARED

The discussion so far has been by topics rather than countries. Let us finally summarize the situation in the various countries. The USA has the closest to a 'free market' housing system, but this does not mean that there is no public intervention or public expenditure on housing. The US Federal Government has spent huge sums on subsidies to owner-occupation and will spend huge sums on the bail-out of the S&Ls. The USA has some horrendous urban problems; it demonstrates that a 'free market' and a high national income per capita do not necessarily ensure good or affordable housing for a country's poorer citizens. The saving grace of the USA is perhaps the variety of approaches which is possible within the States and communities, and the still vigorous 'pioneering spirit' which leads individuals, local commercial organizations, and local government to get together to initiate and finance community projects, including housing projects. This sort of local self-help has rescued much of the (very small) social rented sector in the USA, and will almost certainly be increasingly needed in Europe.

The German Democratic Republic was at the other end of the spectrum from the USA. Its economic system and housing system are well worth studying, for they were widely admired in intellectual circles in Western Europe. And the system did have some virtues. An egalitarian distribution of income was maintained – except for the 'perks' enjoyed by a small élite. There was no unemployment (except for dissidents), and little crime. In some respects, the citizens of the GDR enjoyed good living conditions – like the animals in a well-kept zoo. The extremely low level of rents meant that there were no problems of housing affordability, but the housing system was fatally flawed because of a catastrophic failure of building maintenance, indifference to consumer demand, and the managerial problems of over-centralization.

What has happened in 'the five new German states' since unification – the mass unemployment, the partial breakdown in 'law and order', the disillusionment and moral vacuum – is not only an economic disaster but also a social tragedy. However, given the sudden collapse of the East German state, and the impossibility of restricting immigration into West Germany, the 'instant unification' pursued by Chancellor Kohl was almost certainly the only feasible policy, even though the budgetary costs were ignored.

Time will tell whether the 'new German states' will converge economically with West Germany, or remain permanently depressed. For the longer term, there are grounds for optimism, not least the existence of a federal system which gives considerable freedom to regional and local government; this should eventually replace the current 'takeover'. Although West Germany has deprived the new States of the right to set up non-profit housing enterprises, it may be possible for them to set up housing enterprises which will function virtually as non-profit enterprises, and which will transform the monolithic system of state housing. The new freedoms have already made it possible for owner-occupiers to improve their deteriorating homes. On the other hand, the combination of high unemployment, a more unequal distribution of income, and higher rents are likely to give rise to problems of housing affordability for many years.

The German Federal Republic experienced a 'housing miracle' in the 1950s and 1960s, based on the encouragement of all types of tenure; it coped less effectively with the (relatively minor) problems of the 1970s and 1980s. Complacency about the housing outlook in the light of an expected fall in population led to the adoption of policies – notably the accelerated deregulation of social housing and the abolition of the legal status of non-profit enterprises – which were beginning to cause problems even before unification. The 'old' Federal Republic has now experienced a large inflow of population, which has worsened the housing situation, and it faces the possibility of further unpredictable inflows, at a time when the stock of social housing is declining. On the positive side, the Germans have always responded well to difficulties.

The Netherlands provides an interesting comparison with the GDR. For most of the post-war period, the Netherlands attempted to achieve the objectives of socialism without the defects of the GDR variety. It pursued a range of policies designed to help poorer households, and expanded its social housing stock to currently 43 per cent of the total – by far the largest figure among the countries studied, apart from the GDR. It also developed a system of comprehensive controls of the housing market. Housing costs were comprehensively subsidized; private landlords were instructed whom they might take as tenants; anyone wishing to reside in a municipality, even if they had bought a house there, had to obtain a residence permit.

This massive public intervention has produced many beneficial results. The quality of housing for all social groups and in all districts is relatively good; there would appear to be very little homelessness; the rent ratios for lower income households seem to be lower than in the German Federal Republic – with which the Netherlands is comparable in terms of physical housing conditions. And yet the Netherlands has housing problems (even if relatively small ones)! Part of the explanation is that public expenditure is being cut and controls relaxed – but this in itself suggests that public policy overreached itself. The post-war policy of holding down *all* housing costs led to an unsustainable increase in public expenditure, while private landlords were discouraged by complex financial regulations and the expansion of the public sector. In the major Dutch cities, the public sector has acquired something close to a monopoly, which has produced a selective population structure and various anomalies. The Netherlands faces the difficult task of allowing more play to market forces, while limiting the potentially harmful impact on poorer households.

In France, as in the Netherlands, there has been a tendency for the state to withdraw from a direct role in housing provision. However, a substantial social rented sector has been maintained, and a variety of programmes initiated to encourage the provision of moderate-rent private rented housing, and to cope with the problems of marginal owner-occupiers. France has problems (which are more racial and social than 'housing problems') but it has pursued a quietly pragmatic housing policy throughout the 1980s, with a more consistent balance of 'state' and 'market' than most of the other countries studied.

Housing policy in the UK – for the past thirty years – has been dominated by party politics to a greater extent than in the other countries, and been characterized by policy revolutions. Doctrinaire legislation on housing and housing-related fields has been devised without reference to the work of expert bodies; been rubber-stamped by a Parliament which has effectively ceased to act as a legislature; proved unsatisfactory in practice, and been repealed by the next government. This system of 'strong government' has been disruptive of housing development and management, especially as it affects low-income households. The 'Thatcherite revolution' of the 1980s introduced some housing policies – e.g. the greater use of housing associations – which had long been the practice in other countries, and stimulated more business-like management in social housing.

These policies, however, were given a political slant which resulted in severe hardship to vulnerable social groups. Nor is their much immediate prospect of an improvement in the supply of affordable social housing. Even the Audit Commission, which was set up to be the scourge of local government, has warned that there will be a shortfall in affordable social housing as a result of the virtual elimination of council building and the transfer of council housing to housing associations (Audit Commission 1992). The Commission estimates that 74,000 new social rented units a year will be needed over the next ten years. Councils are unlikely to build more than 5,000 a year. The Government has set housing associations a target of 57,000 units a year by 1994/5. However, housing association finance could well be cut as a result of the growing problem of the Government's budgetary deficit.

A perhaps equally important development has been the dismantling of elected local government (and of strategic planning and public transport). Local government has been either subordinated to Whitehall or replaced by Government-nominated 'quangos' (Urban Development Corporations etc.), appointed without the 'political balance' which was customary for nominated bodies before 1979. The UK now appears to have moved from an era of party political 'U-turns' into a protracted period of one-party rule, in which political change occurs solely as a result of shifts of power within the ruling party. Whether, within such a constitutional framework, 'Majorism' will prove more favourable for the homeless or potentially homeless than 'Thatcherism' remains to be seen. On the positive side, the need for an entrepreneurial, customer-serving approach by housing organizations, and the harnessing of individual initiative, is today preached by yesterday's Marxists.

## CONCLUSIONS

Let us draw the threads of this chapter – and this book – together, in a dozen potted conclusions.

1. In spite of continued economic growth, the ending of the physical post-war 'housing shortage', and a clear improvement in the housing conditions of most people over the past twenty years, a substantial minority of the population in all the Western countries studied has faced increasing difficulties in paying for housing. In the extreme case, this has involved actual homelessness, which affects at most one or two per cent of the population but increased markedly in the 1980s. In this sense, all the countries studied confront a serious 'new housing shortage'. The 'socialist alternative' in the German Democratic Republic avoided these problems of housing affordability, but at the cost of catastrophic failings in the quality, maintenance and variety of housing.

2. There has been a growing consensus among governments that, in principle, the price of housing should cover the cost of providing it. The economic rationale is that only by knowing what a particular good or service costs is a society able to make intelligent choices concerning the allocation of resources. As a consequence, rents of public and private housing have risen steadily towards the cost level, intensifying in the short run the affordability problems of many low-income households. There is a significant gap between the economic cost of housing and what is generally considered a reasonable percentage of income for low-income households to spend on housing. Making housing affordable to low-income households, therefore, inescapably requires some form of housing subsidy. There is, however, no universal agreement on what precisely constitutes an affordable charge for housing. One method is to set a maximum percentage of income (e.g. 30 per cent, including utilities, in the USA). A more equitable method is to vary the percentage according to the size of family and the income level; the larger the family or the lower the income, the lower the percentage payable for housing.

3. Problems of housing affordability arise basically because people are poor. In the 1980s, in all the Western countries studied – in contrast to the earlier post-war period – the rich became richer and the poor poorer. The USA, although the richest country, had a higher percentage of people in relative poverty than the UK, France, the 'old' German Federal Republic or the Netherlands. There was, however, a sharp increase in the percentage of people in relative poverty in the UK.

4. In the long run, housing costs (in relation to average incomes) have been characterized more by cycles than by a clear upward trend.

There was, however, a rise in the 1970/80s, which hit certain social groups particularly hard. Housing costs for households towards the bottom of the income distribution tended to rise disproportionately in most countries, as a result of cut-backs in social housing subsidies and/or a drying-up of the supply of low-rent private rented housing.

5. Housing subsidies – in particular the tax relief on mortgage interest – often went to the wrong people i.e. the rich rather than the poor. All governments have accepted the long-term objective of providing decent housing and living environments for all households. However, no government has yet been prepared to make housing a right of every citizen, and to adopt a subsidy system that will make decent housing affordable to all households.

6. The rising number of homeless people involved a much wider group of persons than in the past; the main increase was in young people, and numbers of women and children began to appear. Many homeless people need social care on a temporary or permanent basis, as well as affordable housing. Many others, however, are simply unable to afford housing, because of temporary problems; they could be helped to escape from the vicious circle of homelessness at a relatively low cost to the taxpayer, through eviction prevention programmes and more targeted housing subsidies.

7. Experience has shown that low-income households' needs for affordable housing are most effectively met by a policy of promoting a variety of tenures – public, private and non-profit rental housing, together with private and cooperative ownership.

8. There was a tendency in the 1980s to replace producer with consumer housing subsidies, but both have advantages and disadvantages. Consumer housing subsidies (e.g. housing allowances) have proved to be an effective way of making housing more affordable for low-income households. They have, however, several limitations. In some countries, they are not an entitlement of all households. In others, administrative defects limit their effectiveness, or the take-up is low. Moreover, they often cover only part of the gap between prevailing costs and what low-income households can afford. Finally, they have only an indirect effect in increasing the supply of housing. Producer subsidies (e.g.through publicly owned housing) have been an effective means of bringing affordable housing within the reach of low-income households, especially in high-cost areas (e.g. areas of rapid economic growth). Housing shortages can persist in particular districts, or for particular types of households, even when there is no

overall shortage, and producer subsidies lend themselves to being targeted on these districts or households. Their main disadvantage is their high initial cost. An effective attack on affordability problems requires a combination of consumer and producer subsidies, with an element of 'public housing'. It is neither necessary nor desirable for local authorities to have a monopoly of social housing, but – to cater for urgent needs – they need to have effective control over *some* social housing.

9. The increasing tendency, in most countries, to use non-profit enterprises instead of local authorities for the supply of 'social housing' has important managerial advantages, and increases the range of consumer choice. At the same time, it raises problems of accountability and control, especially when the non-profit enterprises become very large.

10. The policy of restricting 'social housing' to very low-income households and/or segregating 'problem' (or minority) households in specific housing developments ('sink estates' in Britain) has created severe social problems. Housing programmes which cater for a range of income groups and household types – with appropriate variations in rents or financial assistance – and which disperse rather than concentrate households characterized by poverty and low living standards, develop a stronger social fabric and are more cost-effective.

11. The use of private rented housing owned by private landlords as social housing has the short-run advantage of making housing available more quickly and more cheaply. Difficulties have, however, been encountered (especially in the USA and the German Federal Republic) in keeping subsidized private housing in committed, long-term, low-income occupancy. Greater stability in the stock of social housing is ensured if a substantial proportion is owned by non-profit enterprises, which are private in character but not profit-making.

12. Many 'housing problems' go beyond the physical structure of the housing. The problems in many districts are those of run-down surroundings, street crime, burglary, vandalism, unemployment, lack of education, and the breakdown of family life. Urban decay in central areas is one of the biggest threats to the affordable housing stock. A significant component in affordable housing strategies in all countries has to be governmental measures to protect and enhance local communities; in particular, increased investment in infrastructure, programmes for housing maintenance, and neighbourhood

programmes for enabling young people to become employable, responsible citizens.

## REFERENCES

Audit Commission (1992) *Developing Local Housing*, London: HMSO.

Building Societies Association (1986) *Housing in Britain*, London, pp. 20, 21.

Carter, C. and John, P. (1992) *A New Accord; Promoting constructive relations between central and local government*, York: Joseph Rowntree Foundation.

Central Statistical Office (1991) *Social Trends 21*, London: HMSO.

—— (1992) *Households below average income 1979–88/9*, London: HMSO.

Economic Commission for Europe (United Nations) *Annual Abstract of Housing Statistics*, Geneva.

*Economist, The* (1990) 'Begging', 18 August, p. 24. (G. Randall, *Homeless and Hungry; A Sign of the Times*, 1990, London).

—— (1991) 'American Survey', 29 June, p. 39, London.

Edinburgh, Duke of (1985) *Inquiry into British Housing; Report*, London: National Federation of Housing Associations.

—— (1990), *Inquiry into British Housing; Report*, London: National Federation of Housing Associations.

Fisher, I. (1933) 'The Debt-deflation Theory of Great Depressions', *Econometrica*, 1.

Friedman, M. (1960) 'The Alleviation of Poverty', *Capitalism and Freedom*, p. 190, Chicago: University of Chicago Press.

Greve, J. and Currie, E. (1990) *Homelessness in Britain*, York: Joseph Rowntree Memorial Trust.

Hallett, G. (1977) *Housing and Land Policies in West Germany and Britain*, London: Macmillan.

—— (1988) *Housing and Land Policies in Europe and the USA; A Comparative Analysis*, London: Routledge.

—— (1991) 'Reflections on Corlan', *Welsh Housing Quarterly*, Autumn.

Hambleton, R. (1988) *Urban Government under Thatcher and Reagan*, SAUS Working Paper 76, Bristol University.

Hamnet, C., Harmer, M. and Williams, P. (1991) *Safe as Houses?* London: Paul Chapman.

Hills, J., Hubert, F., Tomann, H. and Whitehead, C. (1990) 'Shifting subsidy from bricks and mortar to people'; experiences in Britain and West Germany', *Housing Studies*, 5:3 July.

Holmans, A.E. (1992) 'House Prices, Land Prices, the Housing Market, House Purchase Debt and Personal Savings in Britain and other Countries', in Milne, A. (ed.) *The Economics of Housing Markets*, forthcoming London.

—— (1991) 'The 1977 National Housing Policy Review in Retrospect' *Housing Studies*, 6:3 July.

Howenstine, E.J. (1986) *Housing Vouchers; A Comparative International Analysis*, Rutgers University.

Kemp, P. (1984) *The Cost of Chaos*, London: Peter King.

Kleinman, M.P., Eastall, R. and Roberts, E. (1989) *Local Choices or Central*

*Constraints; Modelling Capital Expenditure in Local Authority Housing*, Discussion Paper 23:p.58, Dept of Land Economy, University of Cambridge.

Law Commission (1991) *Transfer of Land; Land Mortgages*, Report 204, London.

McFate, K. (1991) *Poverty, Inequality and the Crisis of Social Policy*, Joint Center for Political and Economic Studies, Washington DC.

Maclennan, D. and Niner, P. (1990) *Inquiry into British Housing; Notes*, p.32, York: J.R. Rowntree Foundation.

Merrett, A.J. and Sykes, A. (1965) *Housing Finance and Development*, London: Longmans.

National Federation of Housing Associations (1991) *Review of Low Cost Homeownership*, London.

Newson, T. and Potter, P. (1985) *Housing Policy in Britain; An information sourcebook*, pp. 69–70, London: Mansell.

OECD (1992) *Empowerment and Privatisation Strategies to Address the Housing-related Needs of Low-income Families*, Environment Directorate; Group on Urban Affairs, Paris.

US Department of Housing and Urban Development (1991) *The President's National Urban Policy Report*, Washington DC.

Webb, S. and Wilcox, S. (1992) *Time for Housing Benefit*, York: Joseph Rowntree Foundation.

Wolman, H.K. (1975) *Housing and Housing Policy in the US and the UK*, p. 105, Lexington: H.C. Heath.

# Index

For Product Safety Concerns and Information please contact our EU
representative  GPSR@taylorandfrancis.com
Taylor & Francis Verlag GmbH, Kaufingerstraße 24, 80331 München, Germany

9 781032 003948